Are We Alone in The Universe?

Volume One

Scientific, theoretical, theological, metaphysical, and fanciful perspectives relating to our place in the cosmos

Philip M. Hudson

Copyright 2022 by Philip M. Hudson.

Published 2022.

Printed in the United States of America.

All rights reserved.

No portion of this book may be reproduced, stored in a retrieval system, or transmitted in any form or by any means – electronic, mechanical, photocopy, recording, scanning, or other – except for brief quotations in critical reviews or articles, without the prior written permission of the author.

ISBN 978-1-957077-10-9

Illustrations - Google Images.

This book may be ordered from online bookstores.

Publishing Services by BookCrafters, Parker, Colorado.
www.bookcrafters.net

Table of Contents

Foreword..1
Introduction..7

Chapters

1. Are We Alone in The Universe?...13
2. Love Letters from God..21
3. Tinker, Tailor, Soldier, Sailor..27
4. The Creation..33
5. Dancing with The Stars...39
6. The Mind of God...71
7. Is Music a Universal Language?..77
8. Habla Usted Inglés?...89
9. Heptapod Logograms and The Finger of The Lord................................99
10. Synaesthesia..107
11. The Universe is a Star Nursery...115
12. If You Could Hie to Kolob..119
13. Me Transmitte Sursum, Caledoni...127

14.	Does God Obey the Speed Limit?...133
15.	I'm a Doctor, Not a Doormat..149
16.	Travel at The Speed of Thought..155
17.	The Fluidity of Time...161
18.	The Q Continuum...169
19.	What We Can Learn From the Q?...189
20.	Is God a Carbon-based Life Form?..205
21.	Is There a Prime Directive?..213
22.	Our Genetic Code Has Been Scattered Across the Universe............................221
23.	The Unknown Possibilities of Existence..229
24.	Let There be Light..235

Appendix One..243
Appendix Two..251
Appendix Three...259
Appendix Four..285
About The Author...295
By The Author..299
Quid Magis Possum Dicire?..305
Consummatum Est...309

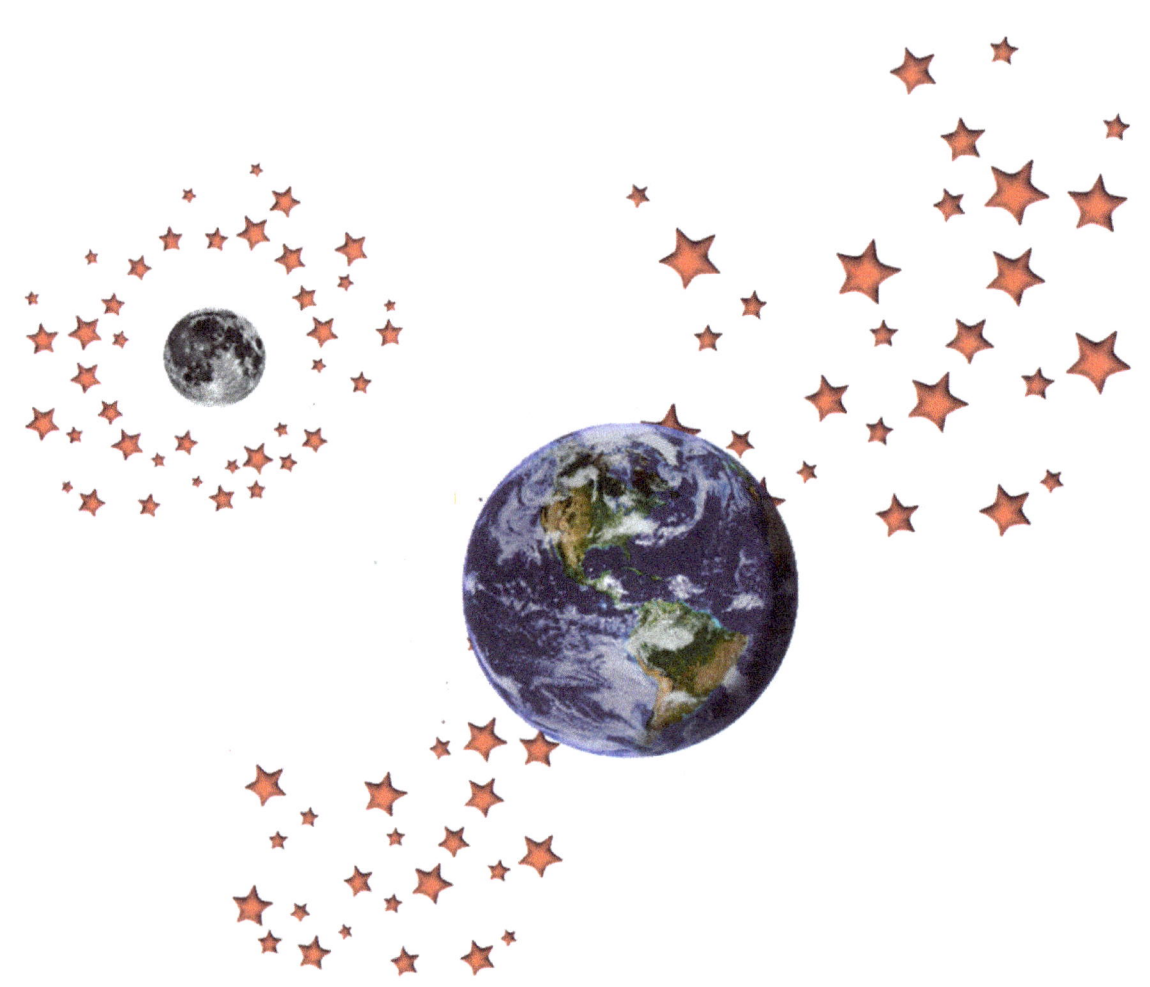

Foreword

Spoiler Alert! This volume squarely addresses a number of straight-forward questions, while offering only elusively vague answers. Instead of closure, its real value is to stimulate further inquiry. When each chapter is studied though the magnifying and clarifying lens of sincere investigation, if we think ourselves full, read ourselves clear, pray ourselves hot, and let ourselves go, nature will caper. We can expect to receive insight, intuition, discernment, inspiration, and revelation pertaining to their subjects.

When we pause in our reading to look up at the night sky, we will ask ourselves again and again if it possible that extra-terrestrial sentient life exists somewhere out there among the stars. A tentative "yes" begs several follow-up questions: Would alien life be familiar, or unimaginably different? Might communication be possible with the inhabitants of those distant worlds? If so, might physical contact also be possible? Would it be hostile, friendly, or indifferent to our overtures? Against the background of religious dogma, what might our relationship be with extra-terrestrial beings? Might we think of them as our long-lost distant cousins? Has communication of some sort already taken place, is it now occurring, and might we reasonably expect it to be commonplace in the future?

Other questions are more esoteric, and address the possibility of communication between ourselves and the glorified inhabitants of even more elusive celestial worlds. Is there a God in heaven Who is the Fashioner and Master of the universe? If so, is communication with Him or Her possible? Might it be possible to establish face-to-face contact? Has First Contact on a level that is spiritually equivalent to subspace communication already taken place, is it now occurring, and if so, can we expect a prayerful or overt dialogue to continue? If so, what would be the conditions of such interactions, and who (or Who) would establish their boundaries?

Related questions probe the depths of our familial relationship with higher dimensional beings from the unseen world. These are metaphysical considerations that lead us to the possibility of moving from the environs of time and space to another plane of existence in eternity. Is such a transition possible? If so, would it be maturational or generational? How can we wrap our finite minds around the temporal, spatial, and dimensional variables that relate to travel to those adjacent or higher realms?

During our armchair travels to the far reaches of our minds as we read this volume, questions that are completely conjectural may pop up to agitate our spirits. They address the possibility of movement from one experiential level of being to higher plateaus within eternal worlds. Is it possible to move through any number of higher dimensional realities within our dynamic universe, or even within parallel universes or between multi-verses? Did our Creator haphazardly or randomly fashion our universe, worlds without end, or did He accomplish it by following a Divine Design? If He had a Plan, might we be privy to His blueprint? Is it reasonable to assume that it is perfect in every detail? Is our galaxy one of billions of star nurseries? Might we be so bold as to think of it as a neonatal incubator and a machine for the making of gods? Have all His children been created in His image and likeness? Is He the Grand Architect of the cosmos? Can we hope to receive divine approbation from the One Who has described Himself as both Alpha and Omega, and the Beginning and the End? As we pursue our dream of making First Contact with our relatives in the cosmos, shouldn't we also expend equivalent, or greater, energy to reconnect with our Celestial Next of Kin?

These questions, and over a hundred others scattered throughout the text, are the rocks and boulders around which the chapters within this volume will turbulently flow. And yet, in between the rapids there will be calm water where we may engage in quiet reflection. We may even find ourselves pinching our noses and taking a leap of faith. If we are nimble, and if we maintain our balance as we hop from rock to rock, and boulder to boulder, we may find them to be stepping-stones that will lead us to a wonderland of independent discovery.

Within each chapter, if you keep looking for them, you will find that their pages are peppered with elements of all these questions. Hopefully, answers will muscle their way to the forefront as you ponder and pray and wrestle for enlightenment, even as other questions germinate in your mind and spring up in the fertile soil of your serious inquiry.

On the flyleaf of his personal Bible, Sir Walter Scott penned the following lines, that apply to holy writ, but that could be interpreted as an oblique reference to profane works such as this one. "Within this awful volume lies the mystery of mysteries. Happiest are they of human race, to whom their God has given grace; to read, to fear, to hope, to pray, to lift the latch, to force the way. But better had they ne'er been born, who read to doubt, or read to scorn."

I hope each chapter in this and its companion volume will become a springboard for your own personal enlightenment, as you probe the mind of God, which is, after all, the final frontier of exploration. May you be introduced to the wonders of exciting new dimensions of experience, as you seek out new relationships and new ways of looking at life, as you catch a glimpse of eternity, and as you visualize yourself living among the stars. May you boldly go into that far country where few have gone before. For your own part, through the ministrations of the Spirit, may you better understand its realm, which is the abode of the Gods.

Introduction

God has instilled within each of us a sense of curiosity that almost compels us to stare in wonder at the night sky, as we attempt to absorb what seems to be an infinite number of stars. The Milky Way is a glowing smear of light across the heavens that is cast from 100 to 400 billion stellar furnaces. It mesmerizes us, and we ask ourselves: "Are we alone in the universe?" To think so begs credulity, and yet, we have discovered no terrestrial evidence of life elsewhere in the cosmos.

Still, myths from around the world give our galaxy its name and explain its origin. The Greeks believed it was created when suckling Heracles dribbled the breast milk of Hera, the wife of Zeus, across the night sky. It was also described as the trail to Mount Olympus, the home of the Gods, and as the path of ruin made by the chariot of the Sun God Helios. In Sanskrit, the Milky Way was called Akash Ganga (Ganges of the Heavens) and was considered sacred. Hindu cosmology explains the galaxy as an ocean of milk churned by the gods for a thousand years, to finally release Amrita, the nectar of immortal life. Heavenly Father put our knowledge of the Milky Way and the universe, estimated to contain between 200 billion and 1 trillion similar galaxies, in a divine perspective, when He told Moses: "The heavens, they are many, and they cannot be numbered unto man; but they are numbered unto me, for they are mine." (Moses 1:37).

Although a luminous glow of light pollution hides the Milky Way from nearly 80% of those in North America, and from nearly 1/3 of the population of the world, it continues to hold us in its grasp with an almost mystical power. It dazzles our minds and figuratively illuminates every corner of our spirits.

Particle physics tells us that, at the moment of the Big Bang and the creation of the universe, there was a release of an incomprehensible number of photons, which are the basic units of electromagnetic energy. The number is 1 followed by 89 zeroes, and yet, it is essentially insignificant when compared to the intrinsic luminosity of God.

At the end of the day, when we have laid aside our scientific instruments that have been designed to probe the heavens for signs of extra-terrestrial life, we will reach the conclusion that "there is no power but of God." (Romans 13:1). Even the photonic energy that we can define, measure, and quantify, has been created by Him to provide reliably consistent light in a world that might otherwise be dark and dreary. Perhaps this volume will be able to capture a few rays of that heavenly aether.

Peter Pan told Wendy that she could find Neverland by taking the second star to the right and continuing straight on 'til morning. As humankind undertakes the exploration of the cosmos, we might want to heed those directions. Captain James T. Kirk of the Starship Enterprise (NCC-1701) did, after Lt. Uhura received a message from Starfleet Command, informing him that the Enterprise was to proceed immediately to Space Dock for decommissioning. Spock responded that, were he an emotional being, he would tell Starfleet to "go to hell." Instead, Kirk gave the helmsman, Lt. Sulu, a more measured order; to follow the second star to the right, to see where it might take them. (See: "Star Trek: The Undiscovered Country" 1991). When writing this volume, I have attempted to follow his example.

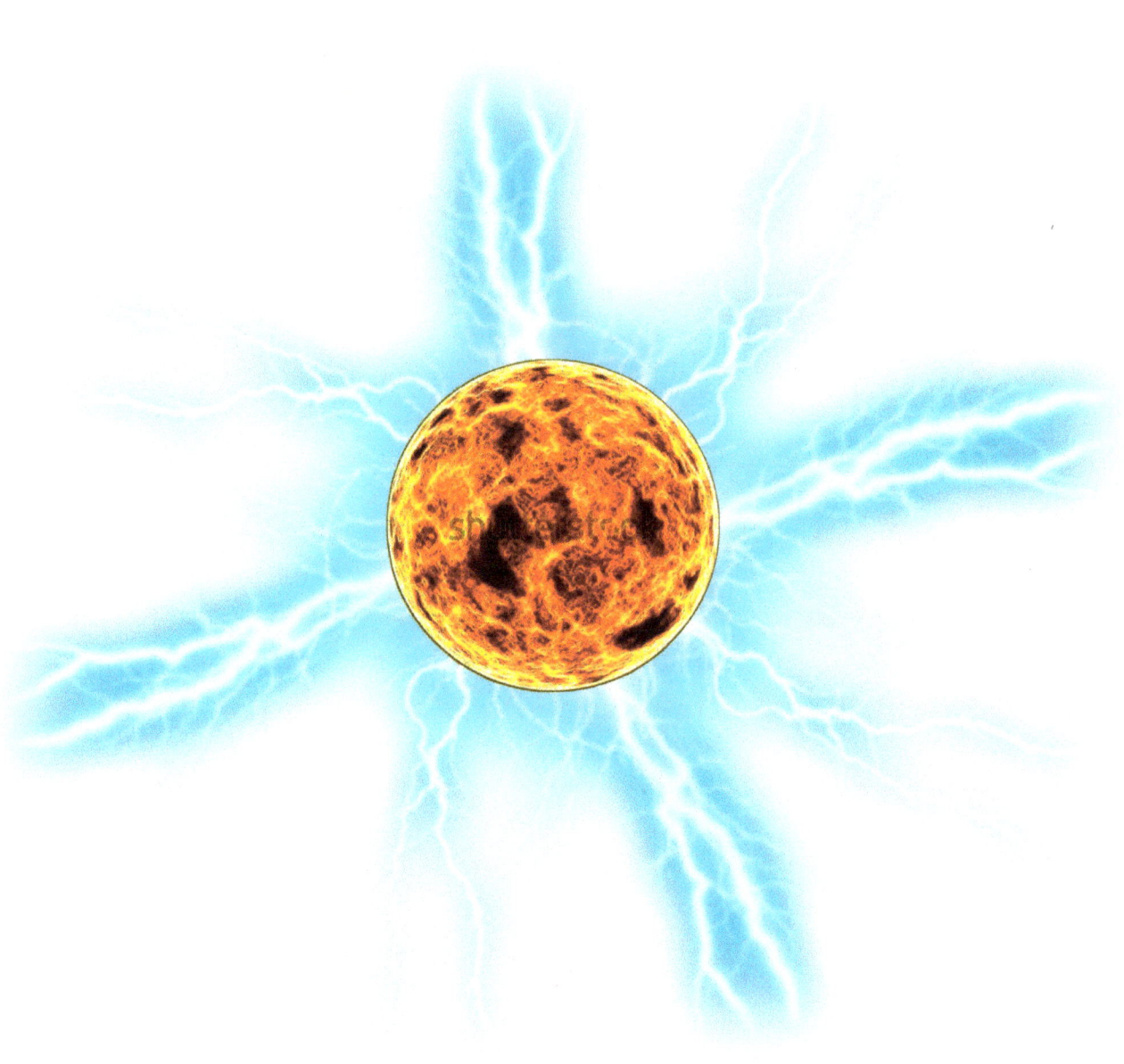

Chapter One

Are We Alone in The Universe?

If you hold up a grain of sand at arm's length against the night sky, hidden behind the blocked-out area there will be roughly 2,000 galaxies, each of which might contain 100 billion to 1 trillion or more stars. The Hubble Space Telescope can "see" these galaxies. It is so powerful that, were it in New York City, it would be able to distinguish between two fireflies in Tokyo that were just 10 feet apart. Expanding that scale exponentially, if each of those 2,000 galaxies hiding behind the grain of sand contains just 100 billion stars, the number of celestial objects within that restricted field of view would be over 200 trillion. That is the number 2, followed by fourteen zeroes, or 200,000,000,000,000, which is, for all practical purposes, "numberless." (See Moses 1:37).

The Hubble is a remarkable $10 billion technological achievement, that has been paying dividends by expanding our understanding of the universe since 1990. For example, in December 1995, the Hubble Deep Field (HDF) captured a photograph that many believe is the most awe-inspiring image ever taken. It is a composite of 342 separate exposures created over a period of 10 days. It's Wide Field and Planetary Camera focused on an area within the Constellation Ursa Major, about 26 arc minutes on a side, which is about one 24 millionth of the sky (an area the size of a tennis ball at a distance of 100 meters). Because the field was so small, nearly 100% of the 3,000 luminous objects that were imprinted upon the image were entire galaxies, apart from only a few foreground stars from the Milky Way.

NASA published the following description of this photograph (you can read more about it on HubbleSite.org). "Telescopes are time machines. When we look out at space, we are looking back in time. The light arriving at Earth from the farthest objects in the universe is light that left those objects billions of years ago. We see them not as they are today, but as they appeared long ago. For 10 straight days in 1995, Hubble stared at a tiny and nearly empty patch of sky, gathering all the light it could, and slowly building a picture. What emerged was the Hubble Deep Field. It revealed galaxies fainter than had ever been seen before. The light from some of these has traveled for 10 billion years to reach us. What we saw was a sliver of the early universe, from long before our Sun was born. It is a three-dimensional picture, and a cosmic core sample. The results are a treasure trove of 3,000 galaxies, large and small, shapely and amorphous, varying in size and color, burning in the depths of space. With the Hubble Deep Field, we reach back nearly to the time when galaxies emerged from the chaos of the Big Bang."

But has the Hubble, or any other manmade terrestrial instrument for that matter, taken us any closer to the resolution of the question: "Are we alone in the universe?" It reaches back in spacetime over 10 billion light years (nearly the "distance" back to the Big Bang itself) and has identified wonders such as a "nondescript" cluster of some 73 galaxies, each containing at least 100 billion stars, found in a tiny section of sky just below the bowl of the Big Dipper. These galaxies are so densely packed that 1,000 trillion stars all fit within an area of sky smaller than the size of a postage stamp.

On a scale so vast, it seems reasonable that, within the bustling cosmic laboratories of an infinite number of star chambers, the building blocks of life would be easily and endlessly created. After all, the laws of physics tell us that every heavy element in our own bodies, the calcium in our bones, and the iron in the hemoglobin of our blood, must have been created during the cataclysmic explosions of supernovae. It is reasonable that, within this seething, gurgling, bubbling, and boiling primordial soup, amino acids would have combined to form the building blocks of DNA that comprise the blueprint of life as we know it.

"The molecules that make up our bodies are traceable to the crucibles that were once the centers of high-mass stars that exploded into the galaxy, seeding pristine gas clouds with the chemistry of life. We are all connected to each other biologically, to the Earth chemically, and to the rest of the universe atomically. We are part of the universe." (Neil deGrasse Tyson). When you look in the mirror, what you see in its reflection is nothing less than a star-child.

At the very least, our efforts to comprehend the universe lead us to a greater understanding of ourselves. When we ask: "What is the origin of the cosmos?" or "Why do its disparate elements behave as they do?" or "What is its ultimate destiny?" or "Are we alone in the universe? what we are really asking is "Where did we come from, and where are we going?" A related question is "Why are we here?" These questions are rooted in the shimmering background radiation from the Big Bang that makes our blood hot to the touch.

In 2021, data from NASA's New Horizons space probe helped to revise its estimate of the number of galaxies in the universe, to roughly 100,000 billion galaxies (1×10^{14}). They surmised that, in a typical galaxy, there are 100,000 million stars, although the Milky Way is estimated to contain roughly 250,000 million stars while Andromeda, our nearest galactic neighbor, 2.5 million light years away, is estimated to contain roughly 1,000 billion stars. Using the most conservative estimate of 100,000 million stars per galaxy, however, there could be 20 billion trillion total stars in the known universe.

If only 1 in a thousand of those stars has planets, that leaves 20 million trillion stars with a solar system. If only 1 in a thousand of those stars that have solar systems has planets that are capable of supporting life, that is to say, has a planet orbiting in the "habitable zone", that is still 20 million billion stars. If only 1 in a thousand of those stars that have solar systems, have planets in the habitable zone that actually support life, that is still 20 thousand billion stars.

If only 1 in a thousand of those stars that have solar systems, have planets in the habitable zone that supports life as we know it, that is still 20 billion stars with civilizations just like our own, with people like you and me, going about their business, looking up at the sky and wondering if they are alone in the universe.

And, if only 1 in a thousand of those stars that have solar systems, that have planets that support life as we know it, have just 1 individual who is going through the same trials as we are, that is still 20 million of God's children who feel just as we do.

If you tried to catalogue these earth doppelgangers at the rate of 1 per second, it would take over 231 years. If you spent just an hour finding out everything you could about each of these worlds and their inhabitants, you would be at the task for over 2 million years. If God spent 6 "days" creating each of these worlds, it would have taken Him around 328 million years to do so. Dumbing it down for us, the Lord simply stated: "And worlds without number have I created; and I also created them for mine own purpose; and by the Son I created them, who is mine Only Begotten." (Moses 1:33).

Are we alone in the universe? In 1977, the United States of America launched the cosmic equivalent of the slogan "Kilroy was here" when it sent Voyager 1, like a bottle in the galactic ocean, out into space. After forty-four years, (as of January 2022), thanks to inertia, it has traveled 14.5 billion miles at a constant velocity of 38,026 miles per hour, or just over 10.5 miles per second. (Go to https://voyager.jpl.nasa.gov/mission/status/ for up-to-date flight data). For a comparative reference, the orbit of Pluto around the Sun is just 2.6 billion miles outside Earth's orbit.

Light, travels through the vacuum of space at exactly 299,792,458 meters (983,571,056 feet) per second. That's about 186,282 miles per second, which is a universal constant known as "c," or light speed. It takes light (or electro-magnetic signals from the Jet Propulsion Laboratory) over 21.5 hours to reach Voyager 1, which is now in interstellar space. In about 40,000 years, the probe will be as close as it's going to get to another star: (AC+79 3888 in the Ophiuchus Constellation.) The Golden Record it carries depicts life on Earth. Electroplated upon its surface is a sample of the isotope uranium-238 that has a half-life of 4.51 billion years. It is possible that, one day in the far-distant future, a civilization that stumbles upon the record may be able to calculate the decay of the isotope to determine the age of the probe, and vector its trajectory, to determine its point of origin. (See Chapter Five: Dancing with The Stars).

Are we alone in the universe? Sci-fi motion pictures with catchy titles like "Close Encounters of The Third Kind," "E.T. The Extra-Terrestrial", "Contact", and "Arrival", stimulate our imagination. In fact, the stated mission of SETI, the Search for Extra Terrestrial Intelligence, is to "explore, understand, and explain the origin, nature, and prevalence of life in the universe." To that end, in rural New Mexico, a "Very Large Array" of independent antennae that are each 82 feet in diameter is positioned along thirteen-mile-long arms in the shape of a Y. In essence, the 27-dish group acts as a single receiver with a 22-mile-wide diameter. Its sole purpose is to detect artificial electromagnetic signals from space, in the hope of resolving the Fermi Paradox that asks: "If there is such a strong statistical probability of technologically advanced extra-terrestrial civilizations, why is there no evidence of them?"

Are we alone in the universe? The mind-boggling distances involved in the search for extra-terrestrial life make it difficult for scientists to answer the question. The nearest stars to Earth (other than our Sun) are the three in the Alpha Centauri system, 4.27 light years away. Light travels about 6 trillion miles in one year. This places the system about 25.62 trillion miles distant. (If it were headed in that direction, Voyager 1 could make the journey from Earth in just 80,376 years, at its current velocity of 38,026 mph).

"We live and move and have our being" on a Pale Blue Dot far out on the Orion Spur of the Milky Way Galaxy, 26,000 light years from its center. (Acts 17:28). If only 1 in a thousand of its 200 billion stars has planets, and if 1 in a thousand of those has planets like Earth, that is still 200,000 indigenous planets with a hospitable atmosphere that could support life as we know it, within our galaxy alone. Perhaps the most convincing sign that intelligent life exists elsewhere in the universe, and even in our own galaxy, is that, apparently, no-one has tried to contact us. That sobering reality may be related to another question that hits closer to home: Is there intelligent life right here on Earth? There is no definitive answer to that question, which is, even now, being hotly debated.

"Are we alone in the universe?" may be the wrong question. When our inquiry is rephrased, and we ask: "Where did we come from? Why are we here, and where are we going?" the power of creation itself is unleashed in our behalf. Our power "to become" is released from the bondage of our ignorance. With a good measure of humility, our arrogance dissipates. With meekness, our reality expands exponentially, and we view the genesis of life in the universe from a refreshingly new perspective.

Joseph Smith rephrased the question "Are we alone in the universe?" when he cried: "O God, where art thou? And where is the pavilion that covereth thy hiding place?" (D&C 121:1). The Lord answered Joseph's prayer from His eternal vantage point as the Guardian of the Galaxy: "If fierce winds become thine enemy; if the heavens gather blackness, and all the elements combine to hedge up the way; and above all, if the very jaws of hell shall gape open the mouth wide after thee, know thou, my son, that all these things shall give thee experience, and shall be for thy good. The Son if Man hath descended below them all. Art thou greater than he?" (D&C 122:7-8).

Long ago, He asked another of His prophets: "Do not I fill heaven and earth?" (Jeremiah 23:24). One wonders how many other times on far-distant shores of the islands of a sea of stars has the Master repeated His reassurance to His children. Perhaps, at least twenty billion times.

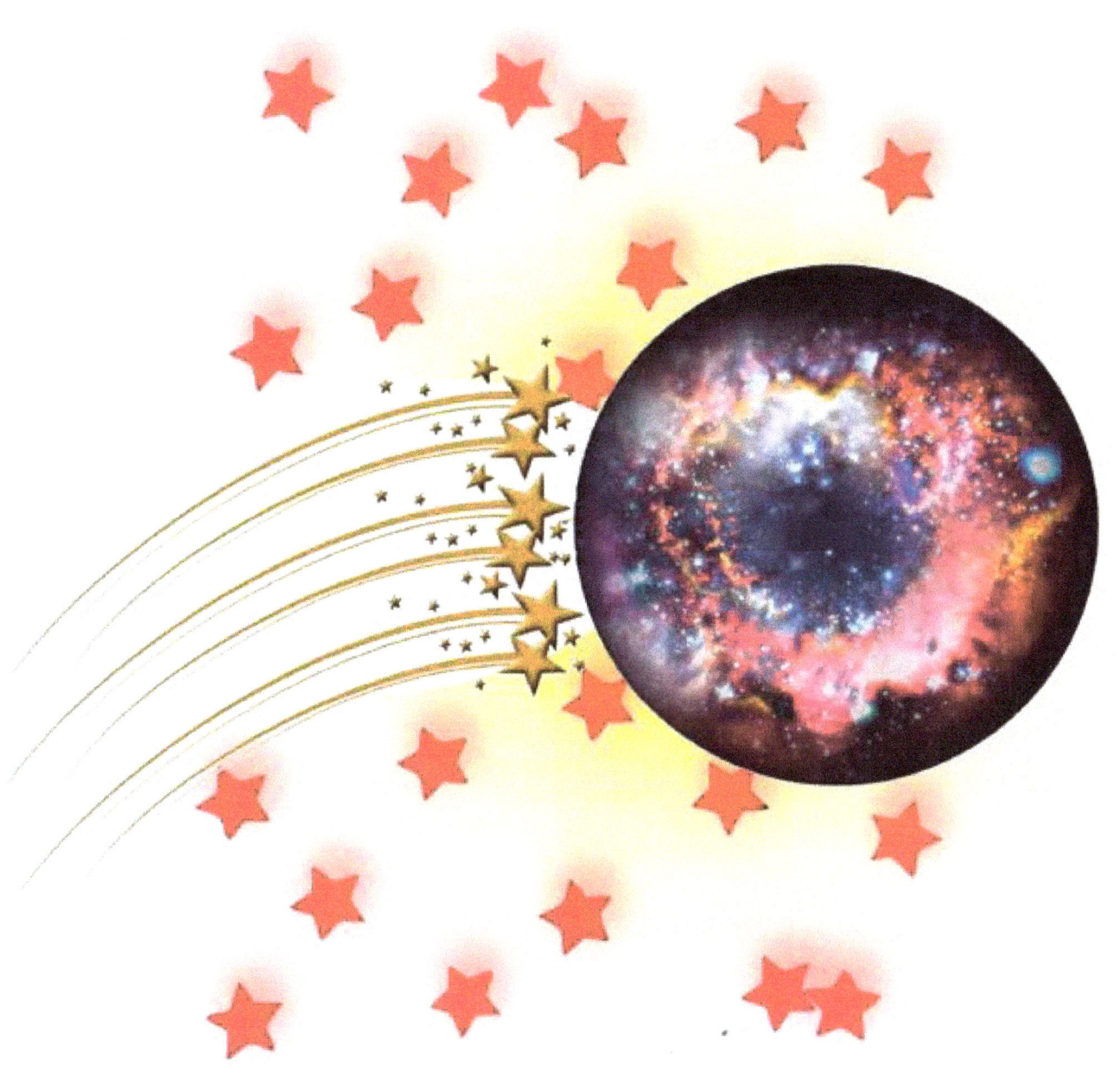

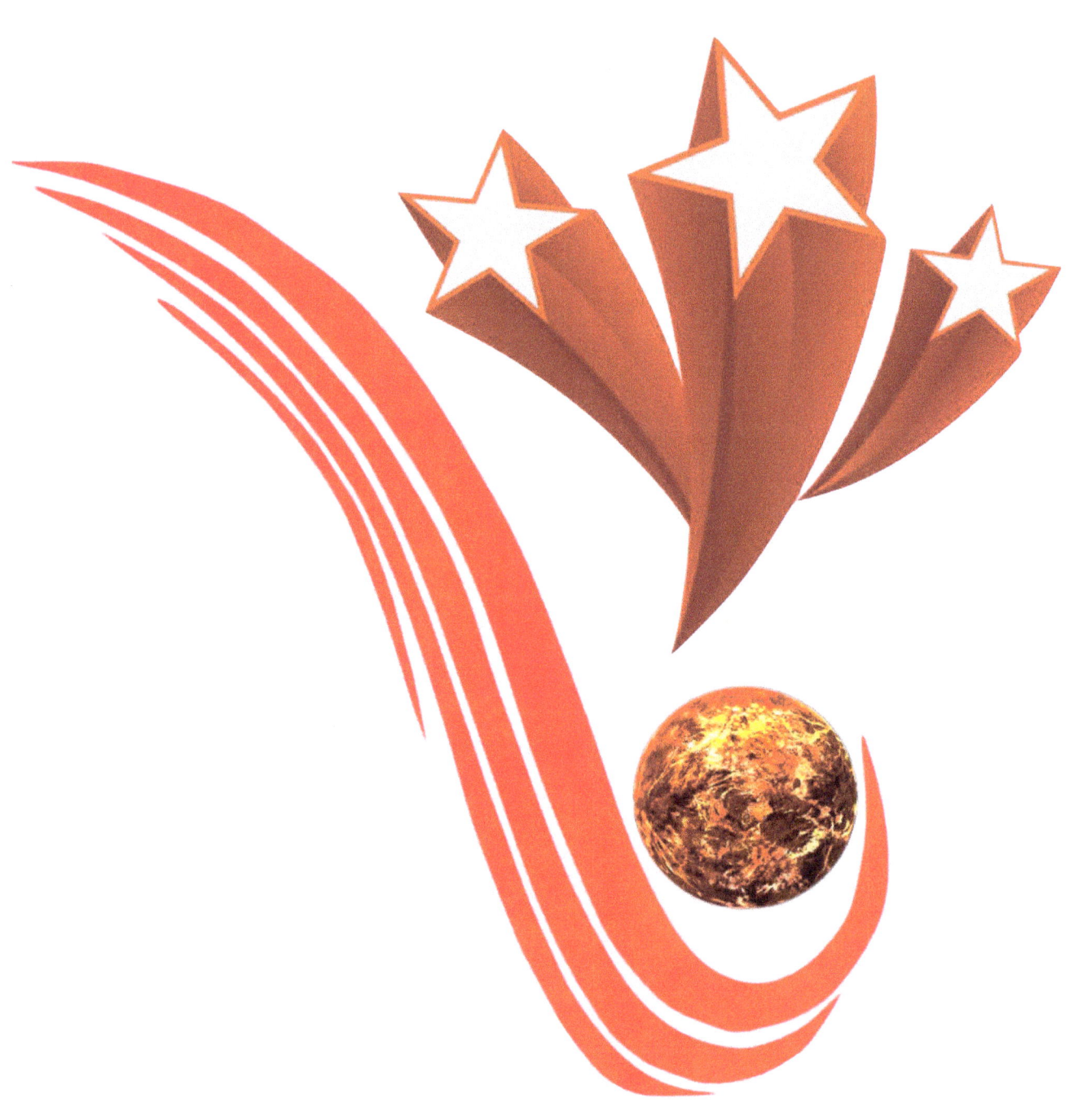

Chapter Two

Love Letters from God

The answer to the question "Are we alone in the universe?" has been revealed in love letters sent from God. "Earth is crammed with heaven," wrote the poet," and every common bush with fire of God. But only those who see take off their shoes. The rest stand around picking blackberries." (E.B. Browning). We don't need a berry bush in our backyard, or a burning bush on Sinai, or smoke and thunder emanating from its summit, to realize that we are not alone in the universe. "The Earth rolls upon her wings, and the sun giveth his light by day, and the moon giveth her light by night, and the stars also give their light, as they roll upon their wings in their glory, in the midst of the power of God. Unto what shall I liken these kingdoms, that ye may understand? Behold, all these are kingdoms and any man who hath seen any or the least of these hath seen God moving in his majesty and power." (D&C 88:42-47).

Ralph Waldo Emerson was particularly sensitive to these subtle expressions of love from God. On one occasion, he wrote: "Those who have seen the rising moon break out of the clouds at midnight, have been present like an archangel at the creation of light, and of the world." On another, he observed: "If the stars should appear but one night in a thousand years, how would we believe and adore, and preserve for many generations, the remembrance of the city of God which had been shown."

From the quiet solitude of his home in Concord, Massachusetts, he mused: "I see the spectacle of morning from the hilltop over against my house, from daybreak to sunrise, with emotions which an angel might share. The long, slender bars of cloud float like fishes in a sea of crimson light. From the Earth, as a shore, I look out into that silent sea. I seem to partake its rapid transformations; the active enchantment reaches my dust, and I dilate and conspire with the morning wind. How does nature deify us with a few and cheap elements!"

With equal sensitivity and insight, William Wordsworth penned the familiar lines: "Heaven lies about us in our infancy! Shades of the prison house begin to close upon the growing boy. But he beholds the light, and whence it flows; he sees it in his joy. The youth, who daily farther from the east must travel, still is nature's priest, and by the vision splendid, is on his way attended. At length, the man perceives it die away, and fade into the light of common day." ("Ode: Intimations of Immortality"). All too frequently, when we secularize the sacred and narrowly view life through the myopic lens of mortality, profound truths are homogenized into easily digestible forms, ennobling principles are compromised, expediency replaces disciplined focus on celestial signposts, and we choke over questions that even babes would understand. "I love everybody, even though some people make mistakes," said five-year-old Kathryn Hudson. "Where did you learn that?" she was asked. "In Church?" "No. When I was up in Heaven. Father told me that." She knew the answer to the questions that vex both doctors and philosophers alike.

Only after we have exerted the effort to squeeze through the strait gate, that we might embark upon the narrow way, will the scales of darkness fall from our eyes and will the way before us open into broad boulevards lined with fig trees laden with fruit, flooded by sunlight, caressed by soothing breezes, and paved with cobblestones that glint of gold. (See Matthew 7:14). Only then will billboards no longer clamor for our attention, will neon lights no longer distract us, and will a cacophony of voices no longer assault us from every direction to suppress the quiet serenity of our comfortable corner of the universe. "We will slip the surly bonds of earth and dance the skies on laughter-silvered wings. We'll climb sunward and join the tumbling mirth of sun-split clouds, and wheel and soar and swing high in the sunlit silence. Hovering there, we'll chase the shouting wind along, through footless halls of air. Up, up the long delirious, burning blue, to top the windswept heights with easy grace, where never lark, or even eagle flew. And, while with silent, lifting mind we tread the high untrespassed sanctity of space, we'll put out our hands and touch the face of God." (Adapted from John G. Magee, Jr., "High Flight").

Are we alone in the universe? We desperately hope that there is a Power Who holds "the key of the mysteries of the kingdom, even the key of the knowledge" of eternal worlds. (D&C 84:19). As we widen our perspective, extend our depth of field, and lift our eyes to strain beyond the limited horizon of our vision, we intuitively know the answer to that question, and we find ourselves cast off into the streams of a revelatory expansion of knowledge. We are carried along in the quickening currents of direct experience with God. We finally appreciate that "the universe is a machine for the making of gods." (Henri Bergson, "Two Sources of Morality and Religion").

Job asked if we are alone in the universe, writing: "Canst thou find out God? Canst thou find out the Almighty?" His habitation "is as high as heaven (and) the measure thereof is longer than the Earth, and broader than the sea." (Job 11:7 & 9). Of our connection to companion powers that surely exist out there among the stars, William W. Phelps wrote: "No man has found pure space, nor seen the outside curtains, where nothing has a place." Nevertheless, there is One Who enjoys "no end to matter, space, spirit, or race, virtue, might, wisdom, or light, union, youth, priesthood, or truth, glory, love, or being." (William W. Phelps, "If You Could Hie to Kolob").

Are we alone in the universe? Are there others who yearn, as we do, to make First Contact? Is their existence defined by the same or different bounds and conditions in the infinite reaches of immortality and eternal life? "I wish I could remember the days before my birth," mused the poet, "and if I knew the Father before I came to earth. In quiet moments when I'm all alone, I close my eyes and try to see my Heavenly home. Where is heaven? Is it very far? I would like to know if it's beyond the brightest star. Where is heaven? Will you show the way? I would like to learn and grow and go there some day. I wish I could remember my Father's loving face, and all the friends and family that filled that holy place. Was I a child there? Did I walk with God? And was that where I learned about the Iron Rod? Although I can't remember and cannot clearly see, I listen to the Spirit and so I must believe. But still I wonder, and I hope to find the answer to the question that is on my mind. Where is Heaven? Is it very far? I would like to know if it's beyond the brightest star." (Janice Kapp Perry, "Where is Heaven?").

If we refuse to acknowledge the certain reality that we are not alone in the universe, that its Master lives out there somewhere beyond the Trail to Mount Olympus, where will our sanctuary be when the winds blow, and the rains beat down? To what safe harbor will we flee when the ocean of life is in turmoil? When we are tossed about as flotsam and jetsam, never coming to a knowledge of what is real, to what source will we look for the stability we so desperately seek, or for the answers to life's greatest questions that continually trouble our spirits? When we raise our sights to the possibility of an expanded view of life, we are up and moving on the pathway to personal re-discovery and self-actualization in the larger arena of higher-dimensional awareness. Perhaps Captain Jean Luc Picard of the Enterprise-D was correct, when he declared: "Space is the final frontier."

The dilation of our temporal reality, and the expansion of our spatial reality, whether or not they are accompanied by First Contact with extra-terrestrial beings, will liberate us from fear, doubt, apprehension of danger, the turmoil of the world, and from the vagaries of men. When we have cast off the self-limiting conditions and self-defeating behaviors that blind us to a larger view of life, we will enjoy a settled conviction of the truth in our minds that affords us the peace that follows obedience to celestial beacons and brings the greater reality of eternal worlds into focus. When we realize that we are not alone in the universe, but that we have a Father in Heaven, we will have begun a journey that will carry us beyond artificial event horizons, to an incomprehensively greater appreciation of the pure love of Christ. (See Moroni 7:47).

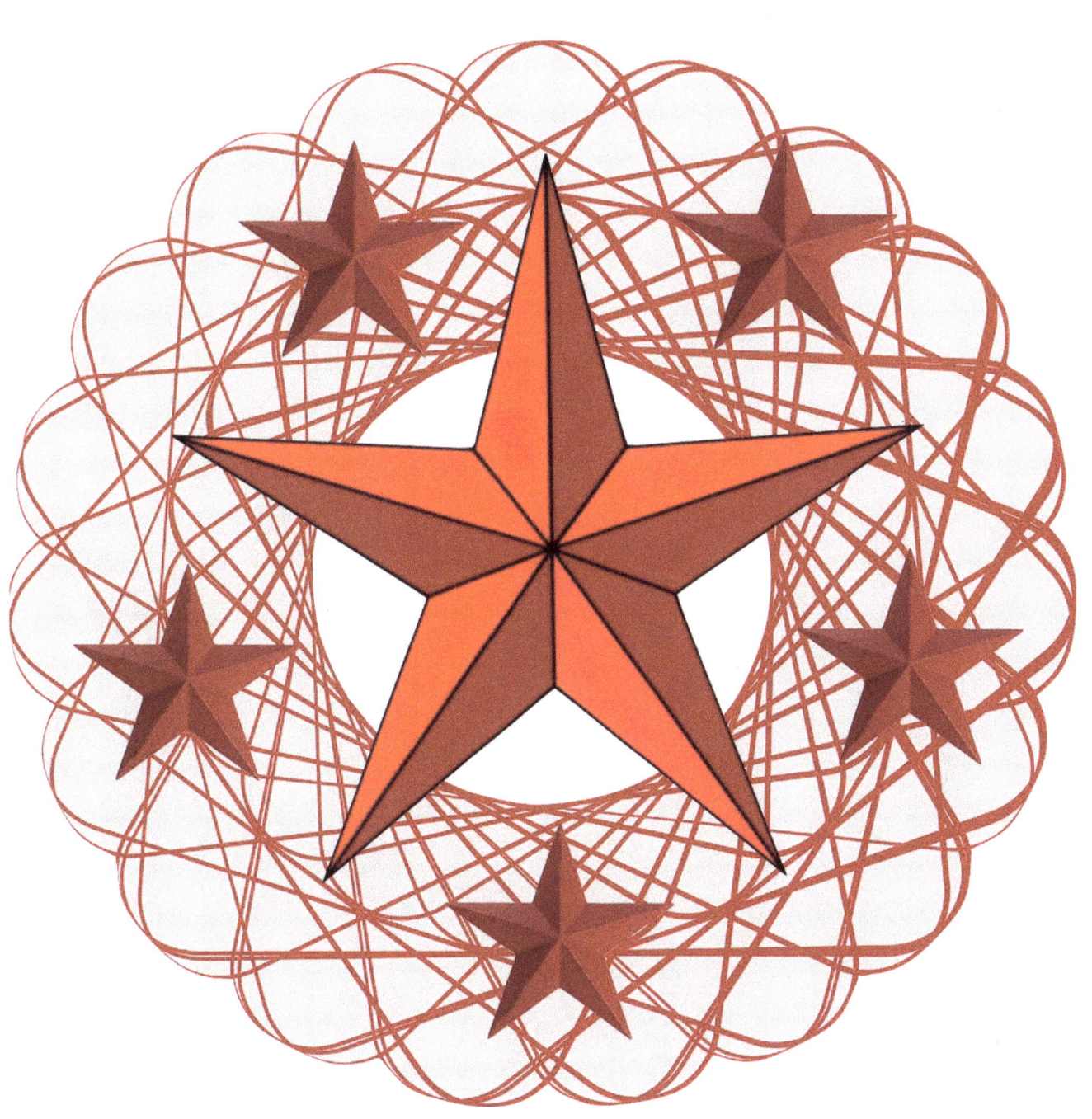

Chapter Three

Tinker, Tailor, Soldier, Sailor

"Tinker, tailor, soldier, sailor." Our understanding of our place in the grand scheme of the universe, where we came from, why we are here, and where we are going, is more profound than the children's numerical counting game would suggest. But those of us with a faith-based perspective of the cosmos face stiff competition from its mathematical rationality, in part because those who pride themselves on its objectivity have often pre-emptively stacked the deck in their favor. They have contaminated the playing field by unilaterally modifying the rules of engagement with their religious counterparts. This creates an obstacle to substantive dialogue and jeopardizes the hope of reconciliation between rational and theological perspectives.

According to deterministic materialism, mathematical constructs originate from our physical interaction with the world around us. In its equations, faith is not a recognized variable, nor are the literal or figurative operations symbols of heaven and hell. Mathematics is a pure science, wherein the fabric of intellectual inquiry is defined by material factors, rather than by the nuance of reasoning to which we are all susceptible, and which, according to the rational mind, is often contaminated by the foolishness of faith.

"Rich man, poor man, beggar man, thief." Mathematical bean counters can be insufferable secular sophists and intellectual guerrillas. They worship at the altars of formulaic logic and the scientific method. In their eyes, only those who think as they do have the intellectual purity to grapple with data and to accurately assess the strengths and weaknesses of argument. Too often, bean counters refuse to concede that there could other ways to understand the world, and the rigidity of rationality lends credence to their suppositions. It is difficult to debate zealots who have been groomed by the Age of Reason, and who have been weaned on the thin pablum that is even now dispensed by the believers of Bacon, the disciples of Descartes, and the supporters of Spinoza.

In his ministry among Athenians who revered Plato, Aristotle, and Socrates as their altar demigods, Paul simply said: "As I passed by, and beheld your devotions, I found an altar with this inscription: To the Unknown God. Whom, therefore, ye ignorantly worship, him I declare unto you." (Acts 17:23-24). The Apostle to the Gentiles, who had been educated at the school of Gamaliel, counted himself not only a minister of Christ, but also a "steward of the mysteries of God." (1 Corinthians 4:1). He declared: "Yet, in the church I had rather speak five words with my understanding, that by my voice I might teach others also, than ten thousand words in an unknown tongue." (1 Corinthians 14:9).

Those five words could have been: "What think ye of Christ?" or "Whose Son might He be?" or "I testify of Jesus Christ." or "I know that He lives." or "You, too, can know Him." or "I will follow my Savior." or "Give up all your sins." They might have effectively been: "I will keep the commandments." or "Cast your burdens on Him." or "Be ye of good cheer." or "Make covenants before holy altars." or "I'm a child of God." or "I'll live with Him forever." Interestingly, there is no surviving record that, in his teachings, Paul ever uttered the words: "Tinker, tailor, soldier, sailor, rich man, poor man, beggar man, thief."

Nevertheless, he employed an economical use of words that would have probably impressed contemporary mathematicians like Pliny the Elder (27 – 79 A.D.) and Hero of Alexandria (10 – 70 A.D.). Paul developed clarity of thought that was as pure as the mathematics of Apollonius of Tyana (15 – 100 A.D.) or Philo of Alexandria (20 B.C. – 50 A.D.). He was concise in conversation, and succinct in his written communication, but clearly he also invited the Spirit to guide his communication. This influenced him to avoid loquacious speech, which gave him an edge with his educated Greek audience, who had been groomed to appreciate erudition in the work of playwrights such as Aeschylus, Sophocles, and Euripides.

We learn in The Book of Mormon how the disciples of Christ emulated Paul. When they prayed to God, they "did not multiply many words." (3 Nephi 19:24). They were continually "drawn out in prayer", but without being verbose, long-winded, rambling, chatty, effusive, vociferous, or garrulous. Oops! (Alma 34:27).

"Tinker, tailor, soldier, sailor." As we rehearse the verbal exchange we hope to one day have with extra-terrestrial beings, we might take a cue from Paul. We hope to be succinct. "Welcome to planet Earth" or "We greet you in peace" or "We mean you no harm." At the same time, we hope that for their part, our alien counterparts will not lead with: "We are Borg. Resistance is futile! Prepare to be assimilated!"

At First Contact, we hope to avoid "profane and vain babblings." We pray that we will not succumb to "fables and endless genealogies, which minister questions." (2 Timothy 2:16 & 1 Timothy 1:4). We have determined beforehand that our speech will be focused and purposeful, and that we will avoid using two words when one will do.

We have begun to send out messages across both the cosmos and the far reaches of cyberspace, and we attentively listen for intelligent responses from somewhere within the Great Silence. Since the adroit manipulation of language might convey the wrong message, we hope to retain the presence of mind to move beyond the mathematical purity, but rational insensitivity, of "tinker, tailor, soldier, sailor, rich man, poor man, beggar man, thief." First Contact will probably require improvisation, as we creatively adapt to the unforeseen circumstances of interstellar and interspecies communication.

Apart from Voyager's messages, and those sent out by SETI's Very Large Array, we anticipate a conversational, and not confrontational, First Contact. We hope that we will be able to sit across from each other, sharing an order of french fries while sipping milkshakes through straws. We hope to avoid small talk, but not make it so loquacious, or so large, that it overloads the capacity of our Universal Translators.

We might take as our cue the Golden Rule, that has 11 words. The Lord's Prayer has 50. The Ten Commandments have 297. The Constitution of the United States of America has 4,500. The Four Gospels, including the words of Jesus, total around 24,600 words. The New Testament contains about 138,000 words. The Book of Mormon has around 268,000 words. But The Affordable Health Care for America Act of 2010 is 1,990 pages long and contains 363,086 words, the IRS Tax Code has about 3,700,000 words, and the U.S. Code has around 42,000,000 words. So much for brevity in official government publications.

"Tinker, tailor, soldier, sailor." We look for a mathematical foundation as we contemplate the existence of extra-terrestrial life among the endless creations of God, and yet, the question: "Are we alone in the universe?" is more than a numbers game. So many galaxies, so many stars, and so many planets, all tossed in a salad with probability as a dressing.

All we know for sure is that God "hath given a law unto all things, by which they move in their times and their seasons; And their courses are fixed, even the courses of the heavens and the earth, which comprehend the earth and all the planets. And they give light to each other in their times and in their seasons, in their minutes, in their hours, in their days, in their weeks, in their months, in their years - all these are one year with God, but not with man. The earth rolls upon her wings, and the sun giveth his light by day, and the moon giveth her light by night, and the stars also give their light, as they roll upon their wings in their glory, in the midst of the power of God. Unto what shall I liken these kingdoms, that ye may understand? Behold, all these are kingdoms, and any man who hath seen any or the least of these hath seed God moving in his majesty and power." (D&C 88:42-47).

"Rich man, poor man, beggar man, thief." The answer to the question: "Are we alone in the universe?" might only be revealed when we move beyond the binary code consisting of so many zeroes and ones that may lie at the foundation of First Contact. It could be that when we hear alien voices for the very first time, they may suspiciously sound like a Southern accent with a twist. One guess is as good as another.

Chapter Four

The Creation

As our circle of knowledge grows, so does the border of darkness that surrounds the kernel of wisdom that we had been so carefully nurturing. The more we know, the more we need to learn. When Moses beheld the creations of God throughout the cosmos, an even greater vision was unfolded to his view. He "beheld also the inhabitants thereof, and there was not a soul which he beheld not; and he discerned them by the Spirit of God; and their numbers were great, even numberless as the sand upon the seashore." (Moses 1:28).

Before we even attempt to understand revealed truth relating to these wonders, we might ask ourselves how much information we would give to a preschool child if we were answering their questions about how an airplane stays in the air, how plants grow, or why we need to breathe. We might provide vague explanations, and only later fill in details that would correspond to their increasing physiological and psychological maturity, academic curiosity, and intellectual capacity.

Even most adults grasp only general concepts without mastering the details that come only through careful scholarship. The same rules apply when considering revealed truth about the Creation. The Lord has given us only a portion of understanding commensurate with our capacity, in our quest to inhale a heavenly ether as we savor the elixir of eternal life. Details may come later, line upon line, when we are more mature in the Gospel, and precept upon precept, when we have a greater capacity and requirement to understand. (See D&C 98:12). For now, we receive revealed truth strictly on a need-to-know basis.

We have yet to formulate a scientific hypothesis that might provide a plausible context for the creation of the 70 sextillion stars within the observable universe. (See Chapter Eleven: The Universe is a Star Nursery). Even with the world's most powerful microscopes and telescopes, we have barely probed what Moses beheld while under the influence of the Spirit. In the meantime, for those who are patiently willing to wait, wisdom and knowledge are continually and seamlessly being deposited by the Holy Ghost upon the bedrock of the foundation of our faith. (See Isaiah 25:9).

The mind of Moses expanded with comprehension when touched by the Lord, and he was able to behold every element of the Earth and appreciate the mathematical and esoteric perfection of God's divine design. His inquiry plumbed the depths of the plasma of the Spirit, and with the powers of celestial cognizance, he began to appreciate the greatness of His creations. As Joseph Smith said: "Could you gaze into heaven five minutes, you would know more than you would by reading all that ever was written on the subject." ("Teachings," p. 156). Thus, if we hope to penetrate the infinite scope of the universe, we must prepare ourselves as have the prophets. Scientific study, though worthwhile, can only contribute a few drops to a bucket that is brimming over with living water from the Source of all truth.

Sir Isaac Newton famously declared: "I do not know what I may appear to the world, but to myself I seem to have been only like a boy playing on the seashore and diverting myself in now and then finding a prettier shell, or a smoother pebble than ordinary, while the great ocean of truth lay all undiscovered before me."

As the Lord God Almighty helped his son to steady his balance on the edge of forever between heaven and earth, Moses might have hoped that a comprehensive understanding of the cosmos would be unfolded to his view. Certainly, he would have been fascinated to know more about God's awe-inspiring creations throughout the universe, and to understand how each piece of the puzzle fit into the Gospel Plan. However, to fulfill his purpose on Earth, this knowledge was evidently unnecessary. He was told: "Only an account of this Earth, and the inhabitants thereof, give I unto you." (Moses 1:35). With only a few oblique exceptions, the scriptures have been silent about the other creations of God. (See Appendix Three, with its 60 + relevant scripture references).

To work out our salvation with fear and trembling before the Lord, it seems clear that He wants us to focus on revealed truth, rather than on unfathomable mysteries whose explanations have been withheld. (See Philippians 2:12). As Alma perceptively told his son: "Now these mysteries are not yet fully made known unto me; therefore, I shall forbear." (Alma 37:11). What has been given to us is a comprehension of that portion of eternal truth that we must believe in order to energize the doctrines of the Fall and the Atonement, that they might become our stepping-stones to the stars. Although an account of the Creation is included in the book of Genesis, its real purposes and importance are expanded upon only in latter-day revelation, as well as in the ordinances of the temple. It is no coincidence that these principles are squarely addressed before holy altars in sacred settings. These doctrines are clearly taught, that we might become heirs of salvation. The Atonement is infinite and eternal, and God is no respecter of persons. (See Acts 10:34). An understanding of how His saving grace works in our behalf may be all that any of His children are obligated to know, wherever they may be living among the stars.

Wilford Woodruff explained: "The Lord Almighty created the Earth that we might come here and exercise our agency. The probation we are called upon to pass through is intended to elevate us so that we can" one day comfortably "dwell in the presence of God our Father." ("The Discourses of Wilford Woodruff," p. 8). Our Pale Blue Dot was created to provide a vigorously healthy environment where we could be tested to see if we would obey God when we were no longer in His presence. It is here that we receive physical bodies, learn to use our agency, gain knowledge, have families, receive ordinances, and make covenants. No less than eight times in the brief account in Genesis of the process of creation, God declared to Those Who assisted Him in His work that it was good. Without a doubt, He knew what He was doing, as He initialized The Plan on our world, as well as on all the others that He had beforehand and has since created as habitations for His children.

Suffice to say that, in the beginning, One among the Gods declared: "We will go down, for there is space there, and we will take of these materials, and we will make an earth whereon these may dwell. And we will prove them herewith, to see if they will do all things whatsoever the Lord their God shall command them." (Abraham 3:24-25). "Herewith" means "with this Earth." In other words, the Lord created the Earth as a learning laboratory, with sufficient room in our mortal curriculum to take both required subjects and elective courses, to accommodate our varying needs and satisfy our God-given curiosity. It would be a place where we could be infused with the positive energy of heaven and proven in tailor-made and uniquely individual circumstances, as we slowly reacquired both the image and likeness of our Father, in a process of reconciliation. (See Abraham 4:26 & 2 Corinthinans 5:18-19). "For behold," said the Lord affirmatively, "this is my work and my glory - to bring to pass the immortality and eternal life of man." (Moses 1:39). By inference, we may conclude that God carries out His work in a similar fashion across the cosmos.

His work was so good, in fact, that it laid the foundations for eternal life, even though "the stars fade away, and the sun himself grow dim with age, and nature sink in years. But (because of The Plan, we) shall flourish in immortal youth, unhurt amidst the war of elements, the wreck of matter, and the crash of worlds." (Joseph Addison, "Cato" Act 5, Scene 1). Among God's numberless creations scattered across an essentially infinite number of galaxies, our experiences on Earth are more than an inconsequential dry run in a forgotten corner of the universe. Similar scenarios have been played out countless times on endless stages that are the settings for the second act of God's Three Act Play. Our mortal experience is not just a tedious dress rehearsal. Every drama in every stellar system in every galaxy throughout the universe is a flawless and integral element of the Great and Merciful Plan of our eternal Creator.

God, Who has taken the time to number the hairs of our heads, fashioned the comfortable home we call "Earth" with us in mind. (See Abraham 3:24, 4:1, and Matthew 10:30). His Plan provided the creative impetus for the Big Bang. Everything that has flowed outward from that point of singularity has been theatrical encore, whose purpose is to ennoble each of us. A little over 4 billion years ago, our particular abode was organized "the same as a man would organize materials and build a ship. Hence, we infer that God had materials to organize the world out of ... chaotic matter" as it existed nine billion years after it had been scattered across the far reaches of the universe. (Joseph Smith, "Teachings" pp. 350).

This explanation has a nice ring to it, because it fits in neatly with the star-child concept of supernova "creation" that was mentioned in Chapter One – Are We Alone in the Universe?" God seeded pristine gas clouds with the chemistry of life, and "the molecules that (now) make up our bodies are traceable to the crucibles that were once the centers of high mass stars. We are connected to each other biologically, to the Earth chemically, and to the rest of the cosmos atomically. We are part of the universe." (Neil deGrasse Tyson). Our blood, and perhaps that of our galactic brothers and sisters whom we will one day meet, runs hot because of the microwave background radiation that is a tangible reminder and remnant of Creation. Even now, our minds similarly expand from a point of singularity to encompass worlds without number. It is as if God has crafted our own personal Big Bang moment. Truly, He is great and ... the number of His years (cannot) be searched out." (Job 36:26).

Chapter Five

Dancing with The Stars

*"You're not alone, you
know ... I said we'd be watching
you, and we have been – hoping that your
ape-like race would demonstrate some growth,
and give some indication that your minds
had room for expansion ... If you're very
lucky, I'll drop by to say hello from
time to time. See you out there."
(Q – Star Trek TNG).*

"The Great Silence" is the contradiction between the astronomically high estimate of the probability of extra-terrestrial life and its corresponding lack of evidence. After all, we ask ourselves: Hasn't humanity reached the point where it should be Dancing with The Stars? Shouldn't we be holding up the Mirror Ball Trophy with our extra-terrestrial quickstep partners? The universe is around 13.7 billion years old and contains something on the order of 70 sextillion (7×10^{22}) stars, many of which undoubtedly have planets, so somewhere, sometime, life should have evolved into societies of technologically advanced species who have either intentionally or inadvertently broadcast their electromagnetic fingerprints across the far reaches of space. Yet, back in 1950, the physicist Enrico Fermi wondered aloud why no such evidence has been detected. His question ("Where is everybody?") has come to be known as "The Fermi Paradox." It is also known as "Silentium Universi." Basically, the paradox is a conflict between scale and probability on the one hand, and the aforementioned lack of confirming evidence on the other. The scale involved is mind-boggling. There are an estimated 200 to 400 billion stars in the Milky Way Galaxy alone. Somewhere out there, intelligent life should be capable of making its presence known. Just think of the electromagnetic signals from Earth alone– for example, all those broadcasts of "Sesame Street" - that have been traveling at the speed of light outward through space since the early years of the 20th century.

The second foundation of the Fermi Paradox has to do with probability; the assumption that extra-terrestrials would have developed the ability to overcome scarcity and colonize new habitat, and ultimately to possess the technology to explore neighboring star systems. But after having had so much time (+/- 13.7 billion years) in which to leave evidence of its existence, none seems to now exist, thereby creating a conflict begging for resolution.

It may be that life is precious, after all, and is found relatively infrequently in the cosmos. Perhaps, we are unique. Hamlet may have been right when he exclaimed: "What a piece of work is man, how noble in reason, how infinite in faculties, in form and moving how express and admirable, in action how like an angel, in apprehension how like a god! The beauty of the world, the paragon of animals." (Shakespeare, "Hamlet," Act 2, Scene 3).

It may be that our own best efforts to reach out and touch someone, anyone, might be corrupted by an "Observer Effect" that significantly alters the perceived state of the elusive objects of our investigation and makes their detection and measurement more difficult. Perhaps, the very devices we have constructed to search for the evidence of alien life have tainted the data that would have otherwise confirmed its existence.

It may be that our neighbors in the infinite reaches of space and time are defined by bounds and conditions that make their detection with our five physical senses impossible. William W. Phelps mused: If only we "could hie to Kolob in the twinkling of an eye, and then continue onward with that same speed to fly, do you think that (we) could ever, through all eternity, find out the generation where Gods began to be, or see the grand beginning, where space did not extend, or view the last creation, where Gods and matter end? Methinks the Spirit whispers, 'No man has found pure space, nor seen the outside curtains where nothing has a place.' The works of God continue, and worlds and lives abound. Improvement and progression have one eternal round. There is no end to matter; there is no end to space; there is no end to spirit; there is no end to race." ("If You Could Hie to Kolob").

Our arguable Shakespearian "nobility" notwithstanding, after 4.5 billion years of evolutionary development, it is still up in the air whether "intelligent" life on Earth will make it past a Type 0 civilization. (See the definition of Types 0, 1, 2, and 3 civilizations, below, and in "Another Perspective" following this chapter). Perhaps, other cultures have embraced, as we have, the insane policy of Mutually Assured Destruction (MAD), that has inexorably led to their ultimate annihilation. Perhaps, the control of exponentially expanding technology proves to be impossible by those moving too quickly along its hopeful path toward Type 1 planetary stability. Perhaps, alien civilizations too often seal their fate by inadvertently opening a Pandora's Box of uncontrollable knowledge and artificial intelligence, thereby sowing the seeds of their own ruin.

Perhaps, burgeoning knowledge harbors a fatal flaw and nurtures a hidden Achilles Heel. As Type 0 societies evolve, though they may envision a technological shield of protection, they may instead inadvertently pursue the creation of a dispassionate "Sky Net." Our own terrestrial storytellers have already described how such a coldly logical machine could one day take on a deadly "life" of its own to bite the hand that had created it. (See the motion picture: "Terminator 3: Rise of The Machines").

Perhaps, intelligent life in the universe has a very difficult time moving beyond our own Type 0 civilization described by astrophysicist and futurist Michio Kaku, with all the petty jealousies, regional conflicts, social and economic inequalities, and sectarian violence that have plagued us for millennia. Inspired by the work of Soviet physicist Nikolai Kardashev, Dr. Kaku has conceptualized Type 1 planetary civilizations, right out of "Buck Rogers" and Type 2 stellar civilizations, like "Star Trek - The Next Generation." He also envisions Type 3 galactic civilizations on the order of The Empire, in the motion picture "Star Wars".

Dr. Kaku believes that "in about 100 years our Type 0 civilization will become planetary (a Type

1 civilization). We'll be able to harness all the energy output of Earth. We'll play with the weather. The danger period is now (Type 0), because we still have the savagery. We still have the passions. We have sectarian, fundamentalist ideas swirling about, and we also have nuclear, chemical, and biological weapons capable of wiping out life on Earth."

He sees two mutually exclusive trends developing, the one toward a multi-cultural, scientific, tolerant, nurturing, interactive, and interdependent society, with easy access to educational opportunity, meaningful employment, and the satisfaction of temporal needs. The Internet, Facebook, Instagram, Twitter, and other social media, rock and roll, fashion, sports, the European Union, NAFTA, Bitcoin, and even English as an emerging planetary language, are evidence that we are hesitantly inching in that direction. We are witnessing the genesis of a Type 1 economy and a Type 1 culture.

However, we also see its opposite in economic inequality and poverty, organized crime and secret societies, political corruption and terrorism, violence against minorities, cultural xenophobia, drug and alcohol abuse, sectarian conflicts, and religious fanaticism, that are all self-destructive reactions against a Type 1 civilization. Whether we make it beyond Type 0, or not, is still undecided.

Dr. Kaku continues: "Now, in outer space, we look for signs of intelligent life. So far, we find none. Civilizations like Type 1 should be commonplace in the galaxy. Some people assume, therefore, that Type 0 civilizations are rather common, but only a few of them make it to Type 1 because their societies, for the first time in their histories, have the ability to commit planetary suicide. Maybe that is the reason why we don't see evidence of alien life. Maybe they never made it. Maybe one day, when we have starships and visit their worlds, we'll see atmospheres that are irradiated because they had nuclear war, or atmospheres too hot to sustain life because they had a runaway greenhouse effect. Maybe, when we explore the galaxy, we'll see the corpses of Type 0 civilizations that never quite made it to Type 1."

At the cosmic speed limit, (Warp Factor 1 - the speed / distance at which light travels in a second - about 6 trillion miles per year), it would only take 120,000 years, (the blink of an eye on a cosmological scale), for a probe or vessel from a Type 2 or 3 civilization to traverse the Milky Way. Our own solar system is a relative newcomer on the 13.7-billion-year-old intergalactic stage, yet we have no indication that sentient life forms have stopped by Earth at any time during its own 4.5-billion-year history, to leave their footprints or calling cards. Perhaps, they have, and we are asking the wrong questions, or are looking in the wrong places for the unique signatures that have indelibly marked their passing.

However, unless alien physicists have worked out the details to permit travel at warp speed, or even trans-warp speed, where space itself is distorted to shorten the distance between points A and B, the physical evidence of life elsewhere in the galaxy, let alone in the universe, may never manifest itself. (See Chapter Fourteen: Does God Obey the Speed Limit?). And yet, there are promising launch platforms for space-faring visitors, right in our own vicinity. For example, the Alpha Centauri System, (our nearest stellar neighbor), is only 4.35 light years away, almost within shouting distance, at 25.62 trillion miles from Earth. A potentially habitable planet, named Proxima Centauri, has been discovered orbiting Alpha Centauri A, one of the stars in the binary pair.

In 1977, the United States of America coined the cosmic equivalent of the slogan "Kilroy Was Here" when it sent Voyager 1 out into space like a bottle bobbing upon a deep and boundless galactic ocean. After forty-four years, (as of January 2022) it had traveled 14.485 billion miles at a constant velocity of 38,026 miles per hour, 0.00005% of the speed of light, or just over 10.5 miles per second. At this rate, it will take Voyager 1 about 14,000 years to travel one light year. (To see how far Voyager 1 is from the Earth, in real time, go to Voyager.jpl.nasa.gov/where). Light, traveling at 186,200 miles per second, takes over 21.5 hours to reach the probe that has now passed the Termination Shock of the solar wind and has entered interstellar space. In about 40,000 years, the probe will be as close as it's going to get (about 1.6 light years) to another star: (AC+79 3888 in the Ophiuchus Constellation). Its Golden Record and pictograms describe life on Earth. Electroplated upon its exterior is an ingenious "atomic clock", a sample of uranium-238 with a half-life of 4.468 billion years. If Voyager 1 has enough specific orbital energy to leave the Milky Way, in one billion two hundred eighty-eight million years it will enter intergalactic space. It is possible that, one day in the distant future, an alien species may stumble upon Voyager 1 and determine its age by calculating the decay of the isotope, and vector its location of origin.

The phrase "a long time ago, in a galaxy far, far away" will convey new and poignant meaning. Twentieth-century humans would be to them as a voice crying from the dust, for our sun would have long since become a red giant after running out of its hydrogen fuel. Although the Earth and its life forms will be destroyed in that process about 5 billion years from now, Voyager 1 would nevertheless remain our enduring legacy, representing our quiet whisper and fleeting influence on the intergalactic stage.

If Voyager 1 were headed in the direction of Alpha Centauri, which happens to be the closest of 51 stars within 16 light years of Earth, it would only take 80,376 years to reach it. The Earth is about 28,000 light years from the center of the Milky Way Galaxy, which itself is around 120,000 light years in diameter. Its closest major neighbor is the spiral galaxy Andromeda, composed of over a trillion stars 2.6 million light years from Earth. These incomprehensibly large numbers suggest that intergalactic two-way travel or communication might just take too much time to complete, even if "ridiculous" or "ludicrous" speed, beyond trans-warp, slipstream, and spore drive, were possible. (See the motion picture: "Space Balls").

So where is everybody? Let's forget our intergalactic second cousins for a moment, (we always considered them a strange bunch, anyway) and just concentrate on our more immediate relatives who might be living in the neighborhood. In our own Milky Way, astronomers have found evidence of 160 billion examples of one particularly promising type of red dwarf star, 40% of which (64 billion) are thought to have supposed "M Class" planets, similar to Earth, in the habitable zone of their star systems. If conditions on only one in a thousand of those 64 billion planets have germinated life, there could be 64 million "earths" out there with thriving life forms. If only one in a thousand of those 64 million has sentient life similar to our own, there could be 64,000 "sister" planets in the night sky, embedded within the fuzzy wash of the light of the Milky Way. On many of these, alien cosmologists might even now be gazing up in the chill of the evening, wondering if they, too, are alone in the universe, and at home their wives might be thinking that, once again, their curiosity is going to make them late for dinner.

Myths from around the world and from the dim recesses of memory give the Milky Way its name and explain its origin. The Greeks believed it was created when suckling Heracles dribbled the breast milk of Hera, wife of Zeus, across the night sky. It was also described as the trail to Mount Olympus, the home of the gods, and as the path of ruin made by the chariot of the Sun God Helios. In Sanskrit, the Milky Way was called Akash Ganga (Ganges of the Heavens) and was considered sacred. Hindu cosmology explains the galaxy as an ocean of milk churned by the gods for a thousand years, to release Amrita, the nectar of immortal life.

Using our aforementioned calculations relating to the probability of life in the Milky Way Galaxy as a baseline, we may extrapolate that within the 600 billion galaxies in the known universe there could be as many as 25.6 million trillion (2.56×10^{19}) "earths" with life just like our own. (See calculations that were published in "Astrophysical Journal", based on observations of the Hubble Space Telescope, (Deep Field series), and with NASA's New Horizons spacecraft). That mind-boggling number pales in comparison to the endless possibilities within multiverses, or within parallel universes, that together might comprise everything that exists and that can exist, the entirety of knowable and unknowable space, as well as of time, matter, and energy. But that is the subject of another ambitious essay.

Perhaps, we have found no proof of extra-terrestrial life because we have an over-developed anthropocentric viewpoint. Perhaps, we look too closely for evidence of the kind we ourselves would provide, without realizing that extra-terrestrial life might behave in entirely different ways that make perfect sense to them, but are completely "alien" to our nature, inclination, or experience.

Then, there is the distinct possibility that the evolution of life on alien worlds has followed a very different trajectory than that of the Earth. To use an example with which we are familiar, the 7.5-mile-wide Chicxulub asteroid struck the Yucatan Peninsula 66 million years ago, at about 27,000 miles per hour, leaving a 124-mile-wide scar on the planet's surface. It's impact effectively ended the Cretaceous Period and nullified the age of the dinosaurs. Were it not for that cataclysmic event, we might even now look and think like our velociraptor distant cousins. Perhaps, alien life is so unlike Homo Sapiens that the gulf separating our species is just too great to bridge with comprehensible communication.

Early in 2012, NASA's planet-hunting Kepler spacecraft confirmed the discovery of the first alien world that lies within the habitable zone of its host star, where temperatures would allow liquid H20 to exist. Later finds brought the space telescope's haul to 2,600 planets. According to a bulletin posted on NASA's website: "After nine years in deep space collecting data that revealed our night sky to be filled with billions of hidden planets – more planets even than stars – the Kepler space telescope has been retired. Kepler leaves a legacy of planet discoveries from outside our solar system, many of which could be promising places for life." The exciting thing about many of these planets is that they are potentially habitable alien worlds orbiting stars very much like our own sun.

"As of January 2015, Kepler had found exoplanets in about 440 stellar systems. In November 2013, astronomers estimated, based on Kepler space mission data, that there could be as many as 40 billion rocky, earth-size exoplanets orbiting in the habitable zones of Sun-like stars and red dwarfs within the Milky Way. The nearest such planet may be only 3.7 parsecs (71.04 trillion miles) away, according to the scientists." (Wikipedia). A parsec is the distance to a star whose position relative to Earth shifts by 1 arc-second (1/3,600 of a degree) in the sky after Earth orbits halfway around the sun. A parsec amounts to about 3.26 light-years, or about 19.2 trillion miles. For a current count of the number of exoplanets Keppler has discovered, visit NASA's Exoplanet Exploration Website – Exoplanets.NASA.gov. (As of November 2021, the planetary count is 4,575).

Using the High Accuracy Radial Velocity Planet Searcher (HARPS) spectrograph at the European Southern Observatory in Chile, astronomers have found within our own Orion Spur of the Milky Way nine more similar planets only slightly larger than Earth. The investigators estimate that about 100 such planets lie in the immediate neighborhood of our sun. The "new observations with HARPS suggest that about 40 percent of all red dwarf stars have an 'earth' orbiting in the habitable zone." (Space: On MSNBC.com on 3/28/2012)

It seems plausible that, over billions of years, intelligent life should have flourished on at least some of these "M Class" ("Earth Similarity") planets that likely permeate our galaxy. The technological accomplishments of humans over the past 50 years, or 5 years, or even the last 5 minutes, that might allow our detection of messages from these worlds, beg the question: "Where is everybody?" If intelligent life is out there, why have we detected no evidence of its existence? At the very least, why hasn't its presence revealed itself through the distribution of electromagnetic signals that plausibly should be the unmistakable signature of civilizations with advanced technologies?

Perhaps, we are squinting our eyes and looking through a tiny keyhole into what may be the wrong microscopic portion of the night sky. When we gaze up into its vast expanse, the visible stars (only about 4,500 of them in either hemisphere) are nearly all in the Milky Way. The only galaxies that are discernable to the naked eye are the Large and Small Magellanic Clouds, which are irregular dwarfs, Andromeda, which is 2.5 million light years away, and, at a fraction of the speed of light, is headed toward the Milky Way on a collision course, Triangulum, the Sombrero Galaxy, and Centaurus A, visible in the southern sky, at 11 million light years from Earth. Maybe the evidence of alien life is not only hidden from our optic nerves and occipital lobes but is also beyond the reach of our most sophisticated instruments, as well.

A view of nearly 10,000 galaxies, (ESAHubble.org), that lie within an area that is just one thirty-two millionths of the sky, is called the Hubble Ultra Deep Field. The photo taken by the Space Telescope includes galaxies of various ages, sizes, shapes, and colors. The smallest, reddest galaxies, about 100 in number, are the most distant objects ever photographed. They formed when the early universe was just 800 million years old. The nearest galaxies - the larger, brighter, more well-defined spirals and ellipticals - thrived about 1 billion years ago, when the cosmos was already 13 billion years old.

Perhaps, First Contact has not yet been made because members of the United Federation of Planets unerringly adhere to The Prime Directive, Starfleet's General Order #1, the most prominent guiding principle in the fictional universe of Star Trek. Interestingly, the Prime Directive, which dictates that there must be no interference with the internal development of pre-warp civilizations, is consistent with the historical real-world concept of Westphalian sovereignty, which is a principle of international law that defines the right of states to inherent exclusive and absolute sovereignty over its own territory. This principle is enshrined in the charter of the United Nations.

The Prime Directive stated that civilizations with advanced technologies should not adjust, alter, amend, change, develop, expand, improve, modify, or revise in any way the natural development of an emerging society, wherever it might be found, even if the aid were well-intentioned or kept secret. To do so, the Federation believed, might radically influence its natural evolution. This could be detrimental to the society itself or to its celestial neighbors within the sphere of its expanding power, or it could be beneficial, but the effect would most certainly not be neutral. In any case, the culture's natural progression, and that of others with whom it might come in contact, would be artificially influenced in ways that would be difficult to anticipate and impossible to control. Regardless of the outcome, there would be no turning back. (A related regulation is "The Temporal Prime Directive", which is discussed in Chapter Seventeen: The Fluidity of Time).

So, perhaps, aliens have been stealthily observing our behavior for some time, but have exercised god-like prudence and restraint when they have been tempted to reveal their presence. Perhaps, they are keenly aware of our sense of urgency, but maintain the disciplined reserve that permits us to take necessary baby steps that might one day lead to Warp Drive propulsion capability and interplanetary or interstellar familiarity. (As an aside, it is NASA's goal to achieve Warp capability by the end of this century. See Chapter Fourteen: Does God Obey the Speed Limit?)

Perhaps, aliens ascribe to the philosophy of "milk before meat," and before they make their presence known, they have determined to see how our transition from a "Type 0" to a "Type 1" civilization goes. (See 1 Corinthians 3:2 & Hebrews 5:12). Perhaps, their patience reflects a more accurate timeline for growth and development than our hasty and imprudent demands for interstellar playdates would dictate. Perhaps, they are patiently waiting for us to develop technologies commensurate with their own. They might invite us to join a Federation of space-faring species only after we have demonstrated that we are willing and able to abide by unimpeachable behavioral, moral, and ethical standards.

On the other hand, perhaps, alien ambassadors are among us even now in the disguise of the greatest minds of the 20th and 21st centuries, conducting tutorials in disciplines that range from astrophysics to zoology. After First Contact had been made in the motion picture "Close counters of The Third Kind," a terrestrial scientist was heard to exclaim: "Einstein was right!" To which a colleague replied: "Hell, Einstein was probably one of them!"

Perhaps, aliens are among us even now, but utilize "cloaking devices" that allow their observation posts to remain hidden from our view until we have independently developed Type 2 technology together with the maturity to move with responsibility among the stars in the untrespassed sanctity of space. Cloaking technologies with which we are already familiar include radar absorbing materials, optical camouflage, and minimization of electromagnetic emissions (in the infrared portion of the spectrum) through cooling. A theoretical cloaking technology concerns the utilization of "metamaterials", artificial substances engineered to have properties that may not be found in nature. These would allow cloaked objects to avoid detection by EM radiation.

Perhaps, the natural inquisitiveness of extra-terrestrials has been tempered by their appreciation of simple math and sobering terrestrial statistics that they have extracted and interpreted from our data bases. In the United States, over a span of 40 years, the annual budget of The National Aeronautical and Space Administration (NASA) has been just under $10 billion, while the budget of the Department of Defense (called "The War Department" until 1947) in just one of those years was $680 billion (2011), or 15% of the federal budget and half of all discretionary spending. The budget of the Department of Homeland Security was $43 billion (2011). (The Defense budget in 2022 was $715 billion.) Perhaps, with alarm, extra-terrestrials have seen where our priorities lie. The members of the "World Nuclear Club" spend $1 trillion each year to maintain their arsenals while jockeying for tactical advantages over their fellow members. In contrast, the annual U.S. Budget for the International Space Station is a paltry $2.1 billion.

Perhaps, we cling to a flawed anthropomorphic assumption about the attitudes of other intelligent species. Perhaps, quite simply, a self-imposed "radio silence" is in effect because extra-terrestrials just want to be left alone. Perhaps, galactic masters (Type 3 civilizations) have overcome the ego-centric desire to affirm their prominence with the profane, self-destructive, self-important, and self-aggrandizing demonstrations with which we are all-too familiar. Perhaps, as they have become increasingly self-actualized and have honed to a fine point their capacity to reach their potential. Over time, they may have effectively extinguished the self-defeating need to draw attention to themselves. Perhaps, their unconscious restraint is a highly evolved defense mechanism and technique of self-preservation.

Perhaps, their reticence is a self-deprecating personality trait that has been carefully cultivated through genetic selection. Perhaps, they have achieved a "star power" that transcends the craving for media attention, admiration, adulation, and adoration. Perhaps, their silence is a galactic, self-effacing understatement. Perhaps, their P.I.N. is such a jealously guarded secret that they view extra-terrestrial knowledge of their existence as the ultimate form of identity theft. Perhaps, they have finally learned how to control the invasive harassment by the paparazzi and the exploitation upon which our supermarket tabloids voraciously feed. Perhaps, they relish their anonymity and cherish their privacy. Perhaps, to hide from prying eyes, they have developed the technological equivalent of the dark glasses worn by our own cultural icons to disguise their identities and protect their privacy.

Perhaps, alien geneticists have manipulated the genome of their species to temper the fires attendant to their transition from Type 0 to Type 1, 2, or 3 cultures. But, in the process, perhaps, their zeal to "go where no-one has gone before" has been selectively bred out of them, as well. Perhaps, they are more comfortable just staying at home "for the evening," hunkered down in overstuffed easy chairs with good books to enjoy by the fireside. Perhaps they just don't want to bother themselves with issues that might rear their ugly heads should they leave their domiciles to bring a warm loaf of bread to their celestial neighbors.

Perhaps, they have discovered still waters and green pastures on their own side of the fence, and they have finally found the elusive inner peace that had been right in front of their noses all the while. (See Psalms 23:2). Perhaps, their tranquility has come at the expense of the primal sense of wanderlust with which we are familiar, and that continues to get our own exploratory juices flowing. Perhaps, the excitement that perpetually pushes at the boundaries of our experience has been suppressed or even extinguished in the hearts of our interstellar cousins. Perhaps, when they "MapQuest" or "Google" potential travel itineraries, they delete without interest the earthly attractions that mesmerize us, like the world's largest ball of twine (measuring 41.5 feet in diameter), or the world's largest toilet paper roll (10 feet high and 8.5 feet across, weighing two tons and containing four million sheets), in Branson, Missouri.

That scenario notwithstanding, there is always the possibility that aliens don't need to pack their bags, gas up the Eagle 5 Winnebago, and experience the thrill of a solar wind in their faces on the open interstellar road or intergalactic highway. Perhaps, they travel at the speed of thought, rather than at the speed of light. (See the motion picture "K-Pax" for a creative dramatization of thought-speed). Perhaps, they dismiss the temporal and spatial limitations with which we are familiar with a cursory wave of the hand that initiates energy bursts deep within the cerebral cortical grey matter of their highly developed brains. As omniscient Q told Captain Picard: "The universe has been my backyard." ("Star-Trek: The Next Generation." "Deja Q"). In another episode, he exulted: "We are going to have fun! I'll take you to places no human could ever hope to see." ("Qpid").

Perhaps, with the evolution of intelligent life, electromagnetic waves are no longer generated because alien communication technologies have become more sophisticated than the crude instruments with which we are familiar, that emit radiation (at a range between 3 kilohertz and 300 gigahertz, and at wavelengths between 1 millimeter and 100 kilometers). Perhaps, William W. Phelps was right, when he declared: "No man has found pure space, nor seen the outside curtains where nothing has a place," even though "the works of God continue, and worlds and lives abound" across the vast reaches of the cosmos. ("If You Could Hie to Kolob").

Perhaps, Type 3 civilizations move about freely in a space-time continuum with which we are unfamiliar. Perhaps, they are just as comfortable navigating tesseracts, (four dimensional hypercubes), as we are the three spatial dimensional boundaries of up-down, front-back, and side-to-side. (See the motion picture "Interstellar"). As "One among them that was like unto God" said, before the Earth was formed: "We will go down, for there is space there," thus making a clear distinction between their limitless natural habitation and our confined world of every day that, at least macroscopically, is defined by just one temporal and only three spatial dimensions. (Abraham 3:24).

Perhaps, we are not only looking in the wrong places, but we are also looking at the wrong time. Perhaps, for extra-terrestrials, its arrow moves not just in one (forward) direction, but in two (forward and backward) directions, or even in an infinite number of other directions. (Think, for example, parallel universes, or multiverses). If aliens have learned to manipulate time as well as space, they may have already established lively communication with our past or our future, rather than with our present. Perhaps, as they have fine-tuned the orientation of their messages, they have been intentionally programmed to reach us at more opportune times when we have been, or may yet be, better prepared or equipped to receive them. (See the movie: "Interstellar" and Chapter Sixteen: Travel at The Speed of Thought). Or, perhaps they have not communicated with us in this fashion, because to do so would be a clear violation of the Temporal Prime Directive. (See Chapter Seventeen: The Fluidity of Time).

The concept of "ancient aliens" popularized by The History Channel and books by Erich von Daniken ("Chariots of The Gods," "Twilight of The Gods," and "Signs of The Gods") suggests that within historical texts, myths, and legends, and among archaeological artifacts, there may be evidence of past human - extra-terrestrial contact. Millions believe, or would like to believe, that a monolith has already been found on the moon (see the motion picture: "2001: A Spacy Odyssey") that was created millions of years ago by an alien race known as "The Firstborn," whose intention was to assist in the evolution of humanity. It may be that in the year 2513, a monolith will yet be discovered in Olduvai Gorge, Africa, buried in ancient rock. If so, it may be the first hard evidence that humankind has benefitted from prehistoric evolutionary facilitators. (See "Marvel Comics").

Should there prove to be monoliths in the real world, a face-to-face encounter with our alien counterparts may already be in the works. In anticipation of such a discovery, perhaps we should now be making a determined effort to scatter invitations throughout the galaxy, announcing to alien races that we have proposed an interstellar orientation meeting at a specific point in normal space and time, that lies in our future. If a Type 2 or a Type 3 alien civilization has mastered time-travel, it should be a simple thing for them to make a dramatic entrance at just the appropriate moment, no matter when the summons may have been received. Our contemporary efforts could then be better directed toward finalizing preparations to receive our invited guests at the aforementioned venue, rather than spinning our wheels without gaining any traction by searching for random electro-magnetic R.S.V.P.s from other aliens who might not wish to be interstellar party-crashers. (See Chapter Seventeen – The Fluidity of Time).

Steven Hawking observed: "The quantum theory of gravity has opened up the possibility that there is no boundary to space-time and that there is no need to specify behavior at the boundary. There are no singularities at which the laws of science break down and no edge of space-time at which one must appeal to some new law to set new boundary conditions for space-time. One could say: 'The boundary condition of the universe is that it has no boundary.'" ("A Brief History of Time: From the Big Bang to Black Holes", p. 136). From their perspective in space-time, aliens from a distant future might even now be getting ready for the announced festivities.

Or, perhaps, aliens communicate at energy levels like those that have been defined by String Theory, that are mathematically complex and observationally obtuse. For example, the "Everett Many-Worlds Interpretation" of quantum mechanics, proposed in 1956, states that all the possibilities described by quantum theory simultaneously occur in a multiverse composed of independent parallel universes that are forever hidden from our view by the laws of physics.

Even if our timing is right, and there is a surfeit of alien Type 2 and 3 civilizations, perhaps we are not looking in the right places, and they have been trying in vain to communicate with us all along, but have been doing so within the medium of the elusive dark matter whose existence is gaining traction among physicists, and that has been postulated to fill 73% of what has heretofore been described as "empty space."

Perhaps, since the moment of creation, there have stubbornly persisted points in space-time that are "without form, and void," to the end that there continues to be "darkness" here and there, then and now "upon the face of the deep." (Genesis 1:2). Perhaps, in these places "worlds and lives abound (while) improvement and progression have one eternal round." Perhaps, in the far reaches of the universe, beyond our observational boundaries, "there is no end to (dark) matter, there is no end to space, there is no end to spirit, (and) there is no end to (alien) race." (William W. Phelps, "If You Could Hie to Kolob").

Perhaps, our efforts to detect signs of extra-terrestrial intelligence have been inadvertently calibrated to the wrong scale. For example, environmentally conscious and economically responsible aliens may have chosen to scatter sophisticated but recyclable "nano-probes" throughout the Milky Way, thereby avoiding the label of galactic litterbugs and dodging the costs of interstellar EPA-equivalent Super-Fund cleanup. If that is so, we may not be able to detect them with our primitive sensors that haven't yet been upgraded to an alien standard of High Definition. Or, perhaps we haven't yet developed the tricorder technology that would allow us to perform advanced sensing, recording, and computing functions. Or, we haven't yet invented the appropriate "SIM (Subscriber Identity Module) Card" to be inserted into our existing instruments.

Perhaps, we have been looking on a terrestrial order of magnitude for the unique signatures of alien life, while our efforts would have been better served had we focused our attention at the atomic or sub-atomic level. Perhaps, we should be scanning, with ever-more sophisticated instruments, the celestial section of cosmically calibrated classified ads, for alien calling cards.

What if evidence of alien life, or even evidence of its attempts at communication, is represented by terrestrial patterns of complex biological information vigorously interacting on subtle sub-cellular levels, and we just haven't noticed, or made the connection? After all, it's only been a handful of earth years, since Watson and Crick unraveled the mystery of D.N.A. To reduce its footprint on the environment, alien life may long ago have become atomic or even sub-atomic, and we simply haven't been alerted to its influence because it's flying well beneath our radar. Perhaps, as we continue to investigate the human genome, we will find embedded within its vast matrix of TACG (thymine, adenine, cytosine, and guanine) the blueprint of our interstellar family history. Perhaps, there is a galactic equivalent of "Family Search", "Ancestry.com", GenesReunited", or "LegacyTree", and we just need to initiate our subscription. As the dog Frank, in the motion picture "Men in Black", observed: "You humans are always looking for the spectacular. Look for something very small, like a jewel."

Carl Sagan has suggested that our natural senses may currently allow us to observe only one of "an infinite hierarchy of universes, so that an elementary particle, such as an electron, in our universe, would, if penetrated, reveal itself to be an entire closed system. Within it, organized into the local equivalent of galaxies and smaller structures, there may be an immense number of other, much tinier elementary particles, which are themselves universes at the next level, and so on forever, an infinite downward regression of endless universes within universes. And upward as well. Our familiar universe of galaxies and stars, planets and people, would be a single elementary particle in the next universe up, the first step of another infinite progression."

"This is the only idea I know," he said, "that surpasses the endless number of infinitely old cycling universes in Hindu cosmology. What would those other universes be like? Would they be built upon different laws of physics? Would they have stars and galaxies, and worlds, or something quite different? Might they be compatible with some unimaginably different form of life? To enter them, poised at the edge of forever," we would jump off into a protoplasmic reality that could be more revealing than any we had ever before experienced, or even envisioned. ("Cosmos," p. 262-267).

If we were to look more carefully, we might discover that alien cartographers have already provided us with the gift of hidden "Easter Eggs" intended to one day catch our attention and nudge us toward a "Stargate" that leads to the heavens. (See the motion picture "Stargate SG1"). If we listen carefully, "we can hear their message even now, as our heads buzz with a hum that won't go away." Perhaps, "our stairway lies on the whispering wind," waiting for us to ascend on a double helix to a new Aquarian Age. (Led Zeppelin, "Stairway to Heaven"). Perhaps, there is "a secret something" that whispers, "You're a stranger here", and you have "wandered from a more exalted sphere." (Eliza R. Snow, "O My Father"). The incontrovertible evidence of our epic journey may have been inserted within the molecular bonds of the chemistry of life itself.

If so, then alien intelligence is more impressive than we could have ever imagined, because it has already subtly communicated with us by planting the seeds of evolution into our DNA. If that is the case, its indelible and enduring stamp has been left upon our nature. Perhaps, the protein-rich primordial broth agitated by the volcanic contortions and seismic contractions of the early Earth was a bubbling alphabet soup created by Intelligent Design to spell out words like potential, innovation, progress, strategy, and success. Perhaps, all along, the evidence we are looking for has been right before our noses. A superior intelligence may be the driving force behind the creative process of our growth and development, subtly working through genetics to provide the push we need to reach for the stars (or even to reach out to heaven).

The Discovery Institute defines "Intelligent Design" as "'certain features of the universe and of living things that are best explained by an intelligent cause, not an undirected process such as natural selection." Think of it. The perennially popular cookbook "The Joy of Cooking" could not have been created without an author. Its existence presupposes a chéf working in the kitchen. The galactic equivalent of a perfect soufflé (on its page 137) may be just the hard evidence we have been looking for to validate the existence of extra-terrestrial Intelligence.

Along similar lines, interstellar space may have been infused with a culture medium grown by extra-terrestrial exo-biologists. Perhaps, we are just now beginning to discern its presence as quorum sensing, the intuitive decision-making process used by decentralized groups (that could be either millimeters or light years apart) to coordinate behavior. Pervasive and complex chemical communication could be the indelible signature of an alien influence on human behavior, as well as on that of other carbon-based life forms, such as mammals, fish, birds, and insects. The biotechnology of quorum sensing could be the lifeblood and common bond of all the species on Earth. Instinctive and sentient behavior could trace their common origins to a cosmic point of singularity, the equivalent of an intergalactic exclamation point, and a big bang of cultural and evolutionary development. Quantum mechanics, the fundamental theory in physics that provide a description of the physical properties of nature at the scale of atoms and subatomic particles, could lie at the foundation of our rudimentary efforts to explain this phenomenon.

If we could somehow unravel the mysteries surrounding the complex matrix of life teeming in air, on land, and within the deepest oceans, there might lie before us, like an open book, the tangible evidence of a continuity of existence that has no temporal or spatial boundary, but that instead thrives on an energy that is felt throughout the galaxy. Even though the origins of the expression "May the Force be with you." are popularly traceable to imaginary extra-terrestrials, it is still one with which we are all familiar. (See the motion picture: "Star Wars").

Maybe, if we tempered our appetite for the wholesale destruction of entire species and ecosystems, we would notice that a majestic clockwork is at play, and we would sense its celestial calibration. Perhaps, we would then discern the quiet ticking away of the precious minutes of a day that is waning, and pre-emptively intervene to avoid the gathering darkness that is looming on the eastern horizon. Perhaps, we could then discipline ourselves to recognize a harmonic pulse, to better feel the surge of a spiritual essence rhythmically beating throughout nature that washes up upon our own shores in perfect cadence across the cosmos. We would be better prepared to fully participate in, and more positively influence, the ocean of thought in a circle of life that is grandly defined and expansively circumscribed by nothing short of the universe itself.

It could be that evidence of alien life has been around us all the while, but we have been too reoccupied to notice. Perhaps, we have not been able to see the forest for the trees. Perhaps, we need to stand back and take a collective deep breath in order to see more clearly that "there is no end to virtue, there is no end to might, there is no end to wisdom, (and) there is no end to light. There is no end to union, there is no end to youth, there is no end to priesthood, (and) there is no end to truth. There is no end to glory, there is no end to love, (and in fact) there is no end to being" itself in the limitless expanse of a universe that is teeming with life. (William W. Phelps, "If You Could Hie to Kolob").

Perhaps, the neurochemical reactions that lie at the foundation of the ideas that pop into our heads are the hard evidence we seek of alien life, since these thoughts often seem to have lives of their own that have outlived their so-called "creators." Perhaps, our déjà vu moments are more than electromagnetic anomalies. They may be the mirrored reflections of forces that are chronologically correct, although they lie just beyond the event horizon of our comprehension. Maybe, this is why inspiration comes, from time to time, as a whisper from the dust "with clarity and freshness, uncolored and untranslated, (speaking) from within (ourselves) in a language original but inarticulate, heard only with the soul." (Hugh B. Brown). (See "Dancing With the Stars" Chapter Ten: Finding Balance in a Chaotic World).

Perhaps, the "codex" of Type 3 civilizations is plainly represented for all to see, in the Periodic Table, the tabular display of the building blocks of life composed of the 118 naturally occurring chemical elements organized according to their properties. In fact, when we look at the "molecules that make up our bodies, we find that they are traceable to the crucibles that were once the centers of high-mass stars that exploded into the galaxy, seeding pristine gas clouds with the chemistry of life. We are all connected to each other biologically, to the Earth chemically, and to the rest of the universe atomically." (Neil deGrasse Tyson). Perhaps, all along, it is we who have been the missing link, the essential ingredient of an interconnected galactic primordial soup, waiting to be served and enjoyed throughout the cosmos.

At the very least, when we ask: "Where is everybody?" our comprehension is energized to embrace expanding self-awareness. When we ask: "What is the origin of the universe?" or "Why do its disparate elements behave as they do?" or "What is its ultimate destiny?" what we are really asking is "Where did we come from, why are we here, and where are we going?" Perhaps, it is the seething background radiation from the Big Bang itself that makes our blood hot to the touch. Perhaps, the faint whisper of barely discernible communication from the stars lies not only within the constant interstellar temperature of precisely 2.725° Kelvin (- 249.425° C), but also within the steady maintenance of our own internal body temperature at exactly 37.0° Celsius.

Or, it may be that extra-terrestrials have evolved beyond life that is based on the biological functions that warm our blood, to embrace a reality that lies outside the narrow boundaries of our physiology. V'Ger (Voyager 1), that on November 23, 2021, at 3:05 p.m. P.S.T. was 14,452,962,172 miles, or 155.4821 Astronomical Units, from Earth, may one day be discovered and reprogrammed by alien mechanical entities and sent back to its home world to establish two-way communication in a universally understood binary language that reconnects with its creator. (See the motion picture: "Star Trek, The Motion Picture", and Chapter Twenty: Is God a Carbon-based Life Form?).

Perhaps, the genesis of our own terrestrial lives can be found elsewhere, and can be traced to an alien cosmic laboratory, where the human genome was nurtured in a secret garden, later to be transplanted into the fertile soil of a primordial Earth that had been carefully cultivated by its Creator to be welcoming in its environment and pristine in its setting. If so, the hard evidence of extra-terrestrial life may be independently confirmed each time we see our reflection in the mirror or share with our friends the photographs of our children and grandchildren. Maybe the most enduring illustration of aliens interacting with terrestrial beings lies in the visual image of storks dropping bundled-up newborn human babies down chimneys in Medieval Europe.

Where is everybody? There is within each of us the innate yearning to know that we are not alone in the universe. The Search for Extra-terrestrial Intelligence, (SETI), is a worthy endeavor because it represents our determination to raise our sights to the possibility of an expanded view of life. Its efforts represent our desire to be up and moving on the pathway to personal re-discovery. Our blood is stirred when we read the journal of the terrestrial explorer Captain James Cook: "I intend to go, not only father than any man has been before me, but as far as I think it is possible to go." ("Captain James Cook: Explorer, Navigator, and Pioneer," BBC). We also recall the words of another captain named James Tiberius Kirk, who, from the bridge of the Starship Enterprise, declared: "Space is the final frontier". (See the motion picture: "Star Trek"). He personified our yearning to imitate the daring and bravado of Cook, who precariously ventured forth on uncharted oceans, driven not only by the wind, but also by noble purpose, to undertake bold voyages in a ship called "Discovery".

Without the evidence that that there are others in the far reaches of space who, just like us, are navigating the ocean of life, and who are going through trials similar to our own, where will our sanctuaries be when the wind blows, and rain squalls beat down? To what sheltered harbor will we flee when tempests toss us about, and our lives are in turmoil?

When we are thrown to and fro as flotsam and jetsam, never coming to a knowledge of what is true, to what source will we look for the stability underfoot that we so desperately seek. Where will we turn for the answers to the questions that continually trouble our spirits, as we attempt to chart a safe passage through shoals and reefs? When we raise our sextant to an interstellar scale that encompasses the possibility of an expanded view of life, and we consider its remarkable potential to catalyze greatness, we will be up and moving on the pathway to personal re-discovery in the larger arena of higher-dimensional awareness. We will have prepared ourselves to face ionic cyclones, temporal rifts, and subspace vortices.

Nevertheless, we must concede the possibility that our desire to contact extra-terrestrials could be dangerous. (See Chapter Seven: Is Music a Universal Language?). As Q warned Captain Jean Luc Picard: "You judge yourselves against the pitiful adversaries you've encountered so far - the Romulans and the Klingons. They're nothing compared to what's waiting. You are about to move into areas of the galaxy containing wonders more incredible than you can possibly imagine, and terrors to freeze your soul." ("Q Who?"). Later in the same episode, after a particularly traumatic encounter with the recently discovered malevolent Borg, Q warned Picard: "If you can't take a little bloody nose, maybe you ought to go back home and crawl under your bed. It's not safe out here. It's wondrous, with treasures to satiate desires both subtle and gross. But it's not for the timid."

It is also possible that we are inadvertently sending the wrong messages, and alien life has chosen to ignore us because of its basic goodness and instinct for self-preservation, when measured against our primitive incivility and shocking lasciviousness. Our electromagnetic signatures could be perceived as threatening, or simply obscene. Think of all those episodes of "Miami Vice" that cast our society in an unflattering light; the motion pictures "Apocalypse Now" and "Full Metal Jacket", not to mention "War of The Worlds", "Aliens", "Predator", and Internet porn. These electronic emissions could be interpreted by our neighbors as caustic and offensive noise pollution on an interstellar magnitude of scale.

To counteract the potentially negative influence, or misrepresentation, of even a minority of these electromagnetic transmissions, on February 4, 2008, at 7:00 p.m. EST, NASA beamed an interstellar dispatch, the Beatles song "Across the Universe", into deep space. It was a message of peace to any extra-terrestrials who might happen to be in the vicinity of Polaris, also called the North Star, in the year 2439. (Polaris is 431 light years distant from Earth). The transmission coincided with the celebration of the 40th anniversary of the song's recording, the 45th anniversary of the Deep Space Network, an international antenna array that supports missions to explore the universe, and the 50th anniversary of NASA itself. "Words are flowing out like endless rain (and) slither wildly as they slip away across the universe. Pools of sorrow (and) waves of joy are drifting through my opened mind, possessing, and caressing me. Images of broken light, which dance before me like a million eyes ... call me on and on across the universe. Thoughts meander like a restless wind ... They tumble blindly as they make their way across the universe. Sounds of laughter (and) shades of life are ringing through my opened ears, inciting, and inviting me. Limitless undying love, which shines around me like a million suns ... calls me on and on across the universe." That positive message had traveled 63.44 trillion miles, as of February 4, 2021. (It still had 2,039.84 trillion miles to go, before reaching Polaris.) Its lyrics, set to music, may represent our best effort to establish a positive connection with extra-terrestrial life.

Such a bond would set in motion an expansion of the appreciation and understanding of our temporal and spatial connectivity with the cosmos. That, in turn, would liberate us from fear and doubt. It would cast off the self-limiting conditions that had heretofore blinded us to a larger view of life. We would expect to enjoy a more settled conviction of the doctrine expressed within an infinitely expanding reality. We would experience the hope of a liberating peace that might follow obedience to mathematically correct physical principles, newly discovered metaphysical mileposts, and spiritually coherent celestial guidelines.

Knowing that we are not alone in the universe would lead to personal epiphanies. We would have begun a journey that would one day carry us beyond every conceivable event horizon to more intensive and reflective self-awareness, deeper and more abiding humility, and ncomprehensively more profound and enduring faith. Our reinvigorated confidence would increase our capacity for mind-expanding higher-level thinking and propel us toward Type 1 planetary stability, and perhaps one day all the way to God's Rest.

When our inquiry ("Where is everybody?") is rephrased, and we ask: "Where did we come from? Why are we here? Where are we going?" the appreciation of creation is unleashed in our behalf, and we experience the exhilaration of personally tailored "Big Bang" moments of self-discovery. It is as if we were present in the V.I.P. viewing section at the moment of singularity itself. Our power "to become" is released from the oppressive bondage of ignorance and from self-defeating behaviors that are the spawn of arrogance. Unprecedented understanding of our genesis creates context, comprehension, and continuity that allows our reality to expand to mind-boggling proportion.

Where is everybody? If we lift our eyes and strain to see beyond our limited horizons, we will intuitively know the answer to that question, and we just might find ourselves cast off into a stream of expanding self-awareness, as we are carried along in quickening currents that take us on a fantastic journey to a far country. We might even come to the point where we begin to appreciate that "the universe is a machine for the making of gods." (Henri Bergson).

As Q told Captain Picard: "Con permiso, Capitán. The hall is rented, the orchestra engaged. It's now time to see if you can dance." ("Q Who?"). Whether or not E.T. is calling, today we are one day closer than we were twenty-four hours ago to the discovery of what it really means to be Dancing with The Stars.

Another Perspective

The Kardashev Rating or Scale measures the technological development of a civilization based on its ability to harness energy. This measurement was proposed by Soviet astronomer Nikolai Kardashev, in 1964. A Type 1 civilization, also called a planetary civilization, has harnessed its planetary energy. A Type 2 civilization, also called a stellar civilization, can use and control energy at the scale of its star system. A Type 3 civilization, also called a galactic civilization, can control energy at the scale of its entire host galaxy.

However, this scale may be archaic, because it focuses its attention on the harnessing of energy, rather than on its manipulation of increasingly smaller quanta of reality, such as creating and destroying matter, and controlling the fabric of space-time itself. For example, in the hypothetical Star Trek Universe, subspace is a concept similar to that of hyperspace. It is a feature of space-time that gets around the cosmic speed limit, permitting faster-than-light communication.

In any event, humanity has not yet reached Type 1 civilization status. Physicist and futurist Michio Kaku suggested that if we increase our capacity for energy consumption at an average rate of 3 percent each year, we may attain Type 1 status in 100 to 200 years, Type 2 status in a few thousand years, and Type 3 status in a hundred thousand to a million years.

As an aside, it is worth noting that Robert Zubrin has adjusted the Kardashev Scale to refer to the presence of civilizations in space, rather than to their energy usage. He termed it the Civilization Range. Within its scope, a Type I civilization has spread across its planet, Type 2 has extensive colonies in its respective stellar system, and Type 3 has colonized its galaxy.

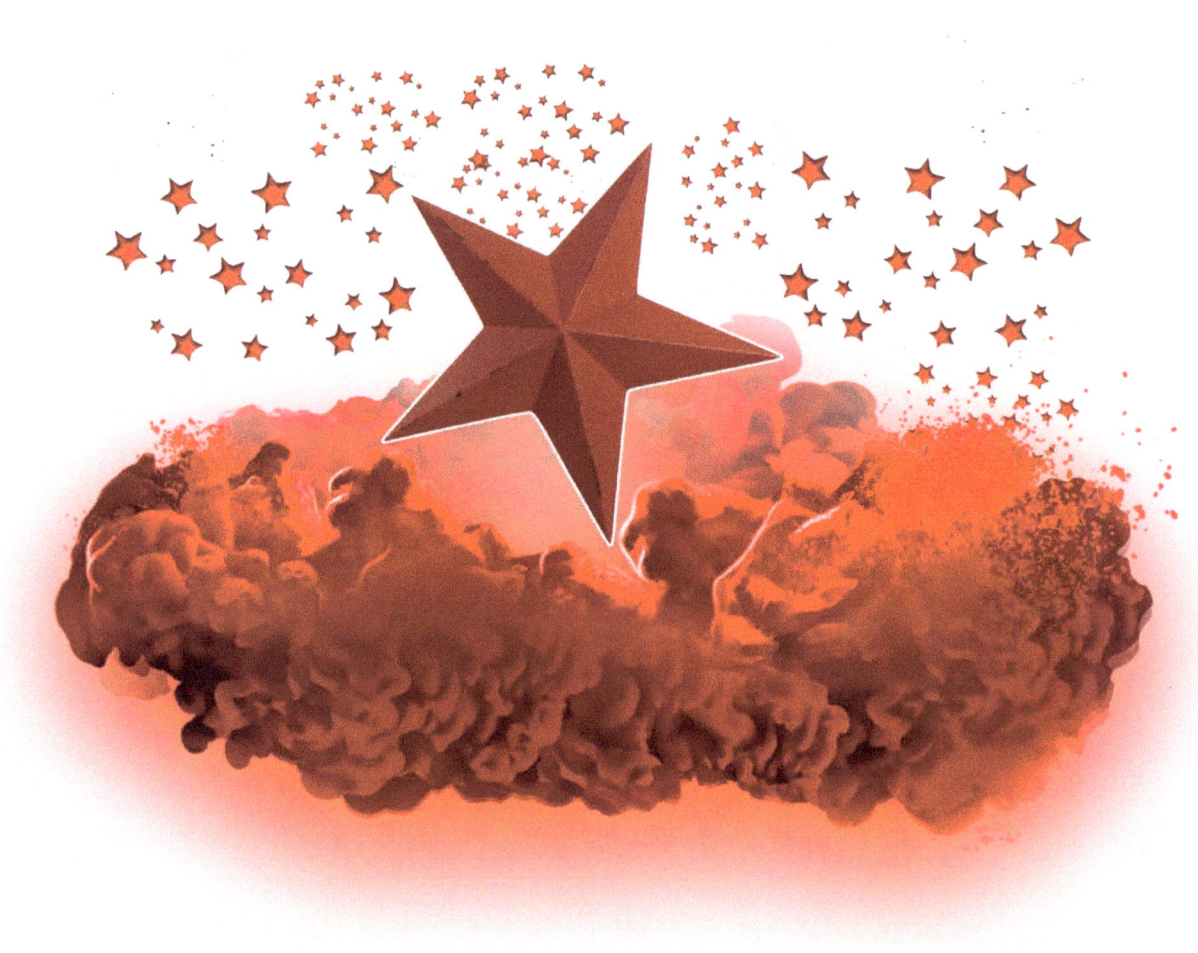

Charting our Course Through the Cosmos

*"A universe simple enough to be understood is too simple
to produce a mind capable of understanding it."*
(John D. Barrow).

*"For as the heavens are higher than the earth, so are my ways higher
than your ways, and my thoughts than your thoughts."*
(Isaiah 55:8-9).

Why haven't we made First Contact with extra-terrestrial beings? Why hasn't the Q Continuum taken an interest in our Pale Blue Dot? Perhaps, their activities are undetectable because they are simply indistinguishable from the everyday workings of nature.

What if we were to extend the Kardashev Scale to include Type 4 civilizations that can control the entire universe by manipulating extra-galactic energy sources such as dark matter, and even to Type 5 beings who control collections of universes, or multi-verses? Borrowing from Star Trek TNG, we immediately think of Q, who invited Captain Jean Luc Picard to actively participate in the evolution of our species. "You just don't get it, do you, Jean-Luc?" asked Q. "The trial never ends. We wanted to see if you had the ability to expand your mind to new horizons, and for one brief moment, you did. For that one fraction of a second, you were open to options you had never considered. That is the exploration that awaits you. Not mapping stars and studying nebula, but charting the unknown possibilities of existence."

As we ponder the potential of a Type 4 or a Type 5 civilization, we again borrow, for perspective, from an exchange between Q and Picard. Picard asks: "Q, what is going on?" Q responds: "You're dead. This is the afterlife, and I'm God." "You are not God," says Picard. Q exclaims: "Blasphemy! You're lucky I don't cast you out, or smite you, or something. The bottom line is, your life ended about five minutes ago, under the inept ministrations of Dr. Beverly Crusher." Picard retorts: "No, I am not dead, because I refuse to believe that the afterlife is run by you. The universe is not so badly designed!"

If we accept as a given that there is a Divine Design, and that creation is controlled, not by the Q, but by less mischievous, more benevolent, more comprehensively omniscient, omnipotent, and omnipresent beings, we must concede that earthlings are the craftsmen of a Type 0 civilization, one that does not yet register on the grand architectural design of the Kardashev scale. With a sigh, we realize that we have a lot of potential, and that God has given us ample room for growth.

But what are our capabilities? Potential energy is energy of position, while kinetic energy is energy of motion. We need to position ourselves to be ready to move, as we look forward to what the new dawn might bring. William Wordsworth wrote: "Tell me not, in mournful numbers, life is but an empty dream! For the soul is dead that slumbers, and things are not what they seem.

Life is real; life is earnest; and the grave is not its goal. Dust thou art, to dust returnest, was not spoken of the soul. Not enjoyment, and not sorrow, is our destined end or way. But to act, that each to-morrow finds us farther than to-day. Art is long, and time is fleeting, and our hearts, though stout and brave, still, like muffled drums, are beating funeral marches to the grave. In the world's broad field of battle, in the bivouac of life, be not like dumb, driven cattle! Be a hero in the strife! Trust no future, however pleasant! Let the dead past bury its dead. Act in the living present, heart within, and God o'erhead! Lives of great men all remind us we can make our lives sublime, and, departing, leave behind us footprints on the sands of time; Footprints, that perhaps another, sailing o'er life's solemn main, a forlorn and shipwrecked brother, seeing, shall take heart again. Let us, then, be up and doing with a heart for any fate; still achieving, still pursuing. Learn to labor and to wait." ("A Psalm of Life").

The potential and kinetic energy described by Wordsworth exceeds that of the known universe, which is within a few orders of magnitude of 10^{45} Watts. Civilizations that theoretically control such power must surpass the limits of speculation and may not be possible. Or, they may be undetectable, as their activities might be indistinguishable from the behavior of nature itself. There might be nothing, except for God, with which to compare the two.

Other proposed extensions to the Kardashev Scale use metrics other than pure power usage, as a baseline. One, proposed by Carl Sagan, measures the information available to the civilization. Another, postulated by John D. Barrow, quantifies the manipulation of our environment over increasingly smaller scales in a process of micro-dimensional mastery. Intriguingly, this "reverse classification" extends downward from Type I-Minus to Type Omega-Minus civilizations.

In this scenario, a Type 1 Minus civilization is capable of influencing solids, liquids, gases, and plasma by joining or breaking the nuclear bonds that hold them together. Type 2 Minus civilizations alter the growth and development of living things by manipulating genes, by transplanting or replacing living tissues, and by interpreting and reengineering their genetic code. Type 3 Minus civilizations create new molecules by making or breaking molecular bonds. Type 4 Minus civilizations are true wizards who are the masters of nanotechnologies on the atomic scale. They are able to create artificial life by manipulating the atomic nucleus itself. They toy with the most elementary particles of matter, fermions (quarks, leptons, and antimatter), and bosons (Higgs bosons and gauge bosons), to create organized complexity out of chaos.

Such mastery inexorably leads to a Type Omega Minus civilization capable of manipulating space and time, or the basic fabric of the universe. For example, they might influence the wave functions of atoms, dissociate themselves from consciousness, ascend to new planes of reality, and take other forms that are incomprehensible to the human brain. Again, we think of the Q, or of God Himself. (By the way, we humans currently seem to be somewhere between Type 3 Minus and Type 4 Minus civilizations.) This begs the question: "Are there non-corporeal species out there that are Type Omega Minus, and who are, therefore undetectable to our senses?" And if there are, might it be more appropriate to refer to them, not as members of the Q or some other Continuum, but rather as "Gods"?

By considering this possibility, and, at the same time, by cutting The Grand Poohbah some slack, we view His omnipotence in a new light. He is the Lord our "God, even Jesus Christ the Great I Am, Alpha and Omega, the beginning and the end, the same which looked upon the wide expanse of eternity, and all the seraphic hosts of heaven before the world was made." (D&C 38:1).

"Listen to the voice of the Lord ... whose course is one eternal round, the same today as yesterday, and forever" Who is "the light and the life of the world – a light that shineth in darkness and the darkness comprehendeth it not." (D&C 35:1 & 45:7). He has promised to reveal more unto us in the "here-after". Therefore, we should let what light He has given us suffice for the present. (See D&C 132.66).

Aristotle felt that life imitates art, although that is still a hotly debated topic. (See Chapter Fifteen: I'm a Doctor, Not a Doormat). That leads us to consider a trail-blazing phenomenon of nature described in Star Trek TNG: the Omega Particle. In that imaginary universe, the Omega Particle had the same power output as a warp core. To put that in perspective, in Starfleet, a warp core engine generated 12.75 billion gigawatts of power. (See: TNG: "True Q"). By comparison, in 2018, the total power output of the United States of America was 153 gigawatts. In the Star Trek Universe, then, a single warp core would provide over 83 million times as much power as that of the entire U.S.A.

The Omega Particle leads us to the Quantum Slipstream Drive. In Star Trek, this power source was an advanced form of interstellar propulsion technology allowing starships to accelerate to warp factor 9.999, or 2,083 times the speed of light. The crew of the USS Voyager acquired this technology in 2374, while returning home from the Delta Quadrant of our galaxy. Similar in principle to trans-warp propulsion, it had been the proprietary technology of an alien race assimilated by the Borg and designated by them as "Species 116."

The quantum slipstream drive operated by routing energy from a power source such as a matter / antimatter reaction assembly through the vessel's navigational deflector, which then focused a quantum field, allowing the vessel to penetrate the quantum barrier, and enter a warp bubble. Real-world theoretical physicists wonder if there are yet other sources of power beyond that of even the Quantum Slipstream Drive. The short answer from a theological perspective is: "Yes!" God, Who is both Alpha and Omega, is the source of that indefinable power "What manner of man is this!" asked a witness of the Savior's demonstration of power over the elements. "For he commandeth even the winds and water, and they obey him." (Luke 8:5) And they "were all amazed, insomuch that they questioned among themselves, saying, What is this? What new doctrine is this?" (Mark 1:27).

Then and now, those with faith to believe know that He is a "God of gods, and a Lord of kings, and a revealer of secrets." (Daniel 2:47). He "made the heaven and the earth by (His) great power and stretched out arm, and there is nothing too hard" for Him to accomplish. (Jeremiah 32:17). He "is a great God, and a great King above all gods." (Psalms 95:3). In his Apocalypse, John saw "the seven angels which stood before God (and) there were voices, and thunderings, and lightnings, and an earthquake." (Revelation 8:2 & 5). John was privileged to see beyond Quantum Slipstream Drive technology into eternity, and when he did, "a door was opened" right through sub-space, into "heaven, and" he heard a voice "as it were of a trumpet talking" to him. (Revelation 4:1).

God's "warp core", however, was only warming up. Moses was privileged to see in action His equivalent of the Quantum Slipstream Drive. He recorded that "all the people saw the thunderings and the lightnings, and the noise of the trumpet, and the mountain smoking" when God spoke. (Exodus 20:18). Peter foresaw the day of the Lord, when "the heavens (would) pass away with a great noise, and the elements (would) melt with fervent heat, (and) the earth also and all the works that are therein (would) be burned up" and "the heavens being on fire (would) be dissolved." (2 Peter 3:10 & 12).

As it turns out, God's ever-present realty reflects Type Omega Minus, and we are Type 0, or most optimistically, Type 3 Minus or Type 4 Minus. Wherever we might now comfortably fit on the spectrum, the Gospel remains our schoolmaster. We navigate our temporal reality with the assistance of about 1.4 kilograms, or 3 pounds, of cerebral cortical grey matter, the equivalent of the dilithium crystals of a warp core, allowing us to continue our feeble attempts to wrap our finite minds around God's infinite power. (See Galatians 3:24).

The Gospel Plan is the vehicle that has been ordained to ease our transition from the world of every day into eternity, where the eyes of our understanding will be opened, and we will be re-introduced to the heavenly equivalent of a matter / anti-matter reaction chamber – a society of celestial beings whose mission statement is to bring to pass the immortality and eternal life of star children throughout the universe. (See D&C 76:12 & 19). For now, the total power of sunlight striking Earth's atmosphere each day is roughly 174 petawatts. For all eternity, however, we will be Dancing with The Stars, utilizing the real star power of the untold billions of petawatts that energize celestial worlds.

Chapter Six

The Mind of God

Latter-day Saints who participate in the ordinances of the Gospel understand that there exists a second order of mind. The experiences of the temple, for example, repetitively reinforce the shadow of otherworld experiences and remind us that there are some things that cannot be understood by the uninitiated, for their symbolism can only be spiritually discerned. This symbolism reinforces the concept that we are not alone in the universe, for we are taught that the events played out in the Garden of Eden were only one of a countless number of similar scenarios. The temple endowment confirms the universal applicability of God's mission statement, which is to bring to pass the immortality and eternal life of all His children, no matter upon what spheres in His vast creations they may reside. (See Moses 1:39).

An interesting twist on viewpoint is the observation that we are presently denied the gift of concentrating on more than one thing at a time, and that to be able to do so would be more real and correct than the perspective we have now. Once we consider that possibility, the universe takes on new dimensions and the mind of God takes over our thoughts.

The Restoration of the Gospel revealed the doctrine that God could preoccupy Himself with countless numbers of things. "The heavens they are many, and they cannot be numbered unto man; but they are numbered unto me, for they are mine," said the Lord. (Moses 1:37). Clearly, we are dealing with two orders of mind. "For my thoughts are not your thoughts, neither are your ways my ways, saith the Lord. For as the heavens are higher than the earth, so are ... my thoughts than your thoughts." (Isaiah 55:8-9).

Orson Pratt spoke of the ability to consider many different ideas at the same time, instead of thinking unilaterally by following only one course of reasoning at any given time. If God were to bless us with a sixth, a seventh, or even a fiftieth sense, he wondered, how would we then view the world? Could each of these senses potentially convey as much information as do our somatic senses? Do our five poor senses represent every element of nature, or is there a much broader panorama of new ways to look at the world?

In such a scenario, "knowledge (would) rush in from all quarters; it would come in as does light that flows from the sun, penetrating every part, informing the Spirit, and giving understanding concerning ten thousand things at the same time; and our minds would be capable of receiving and retaining all. Not one object at a time, but a vast multitude might rush before us, (in) the vision of a celestialized soul."

We might be filled "in a moment with the knowledge of worlds more numerous than the sands of the seashore. Our minds might be strengthened in proportion to the amount of information imparted. It is this tabernacle, in its present condition, that prevents us from that more enlarged understanding." (J.D. 2:238-248).

With such greatly expanded powers of observation, we would more clearly relate to the mind of God. If we were able to look in every direction at once, it would almost certainly reveal a more comprehensive appreciation of His creations. After all, it was under such inspiration that Moses looked, and "beheld the world and the ends thereof, and all the children of men which are, and which were created." (Moses 1:8).

We would receive knowledge that would otherwise be denied to us as long as we remained imprisoned within the narrow confines of our mortal bodies. Recall once again the experience of Moses, who, after he was left to himself, "fell to the earth," so completely overcome "that it was for the space of many hours before (he) did again receive his natural strength like unto man." (Moses 1:9-10).

Elder Pratt reasoned that "there must be some faculty or power natural to God and to superior beings, that we are not in possession of in any great part, by which He can look at a great variety of objects at once." It was only under the influence of the Spirit, for example, that past, present, and future generations all came before the Brother of Jared, and there was not a soul that he did not behold. (See Ether 3:25). Here, then, is a new faculty of knowledge, extended in its nature, that is calculated to throw a vast amount of information upon our minds, almost in the twinkling of an eye. We are not in possession of these powers, but if we were, it could potentially make travel through the cosmos a lot easier. (See Chapter Sixteen: Travel at The Speed of Thought).

Elder Pratt continued his discourse by suggesting that celestial beings have the ability to perceive with all parts of their bodies. If we accept as a given that the Spirit is capable of experiencing the sensations of light, why could we not then see in all directions at once? By extension, who is to say that extra-terrestrial beings could not have mastered the ability to do the same?

This concept clarifies the explanation given by the Prophet Joseph Smith, who described his experience while receiving revelation: "My whole body was full of light, and I could see even out at the ends of my fingers and toes." Perhaps, this is also why the barefoot angel Moroni hovered in the air during his visit to Joseph in his bedchamber. (See J.S.H. 1:31). He wanted to be able to see better, even out of his toes. Think of the ramifications of the Lord's declaration: "If your eye be single to my glory, your whole bodies shall be filled with light, and there shall be no darkness in you; and that body which is filled with light comprehendeth all things." (D&C 88:67).

Brigham Young expounded upon the clarifying light of eternity when he said: "I long for the time that a point of my finger or motion of the hand will express every idea without utterance. The eye will not be the only medium through which we see, nor the brain the only means by which we will understand. When our bodies are full of the Holy Ghost, we will be able to see behind ourselves with as much ease, without turning our heads, as we can see before us. If you have not had that experience, you ought to have. It is not the optic nerve alone that gives the knowledge of surrounding objects to our minds." (J.D. 1:70-71).

Parley P. Pratt believed that, under the influence of the Spirit, we could even move from one place to another without a loss of time, by willing ourselves to be there. "There is no apparent limit," he reasoned, "to the speed attainable by the body, when unchained, set free from the elements which now enslave it." ("Key to The Science of Theology," p. 162).

Brigham Young described the mind of God, when he said: "The brightness and glory of the next apartment will be inexpressible. (Beings in that realm shall) move with ease and like lightning. If we want to visit Jerusalem, or this, or that, or the other place, there we are. If we want to behold Jerusalem as it was in the days of the Savior, or if we want to see the Garden of Eden as it was when created, there we are. We may behold the Earth as at the dawn of creation." (J.D. 14:231). Brigham Young described movement through both space and time, or what we would today call the space-time continuum, a term more recently coined by theoretical physicists utilizing a discipline of science that did not exist in his day.

When the Saints dwell with God in glory, what will their state be like? Since His glory "is intelligence, or in other words, light and truth" (D&C 93:36), it must be that the righteous will dwell amidst the fire and smoke that are symbolic of His presence. We can only guess what our resurrected perfected bodies will be like and what our capabilities will be. We only know that the Lord has promised that we will dwell within the powerful influence of His Spirit. But what an experience that will be! He will "make known ... the secrets of (His) will - yea, even those things which eye has not seen, nor ear heard, nor yet entered into the heart of man." (D&C 76:10, see also 1 Corinthians 2:9). Our "wisdom shall be great, and (our) understanding reach to heaven; and before (us) the wisdom of the wise shall perish, and the understanding of the prudent shall come to naught." (D&C 76:9).

Our experiences in the House of the Lord that expose us to God's second order of mind are sacred for the very reason that it is there that we are privileged to take our bearings on eternity. We are invited to see things as they really are, and through the endowment, we venture into the sacred precinct of the mind of God. From among many spiritual manifestations, Joseph Smith had one of his most profound theophanies in the Kirtland Temple. He later recalled: "The veil was taken from our minds, and the eyes of our understanding were opened." In other words, he enjoyed a sensory experience that revealed to him the mind of God that was so profound that it almost defied description. "We saw the Lord," He testified, "standing upon the breastwork of the pulpit, before us; and under his feet was a paved work of pure gold, in color like amber. His eyes were as a flame of fire; the hair of his head was white like the pure snow; his countenance shone above the brightness of the sun, and his voice was as the sound of the rushing of great waters." (D&C 110:1-3).

The Apostle Paul had similar extra-sensory experiences. He wrote that now we only "see through a glass darkly," but then, if our eye be single to His glory, we shall look upon Him who is eternal "face to face." (1 Corinthinans 13:12). We will begin to understand the mind of God. We can only imagine what it will be like in a coming day to look upon the wide expanse of celestial glories and nod our heads in comprehension of His creations. Perhaps, only then will the riches of eternity have meaning for us and will its solemnities rest upon our minds in an expanding awareness. Perhaps, only when the Lord reveals His wonders, will we see things as they really are.

Chapter Seven

Is Music a Universal Language?

In the Alpha Centauri star system, Proxima B is a planet in the habitable zone. Since it's only 4.2 light years from Earth, it's a good candidate for "instant messaging." In 1973, astronomers sent the Arecibo Message, in 2008 the Bebo Message, and in 2017 the Tromso Message. No one has yet answered, but it could be any day now. In contrast, we don't expect a response to the Voyager 1 or 2 probes for eons.

Some astronomers think we may be sending the wrong greeting. Instead of trying to impress our galactic neighbors with the knowledge of how an atom of hydrogen can flip its polarity at 1,420 MHz, maybe wholesome music would convey a better message. In fact, this has already been done by NASA, as described in Chapter Five, (The Beatles song "Across the Universe"), in anticipation of, and to pre-emptively counteract, the potentially negative influence or misrepresentation of even a minority of a plethora of our electro-magnetic transmissions. The music of the NASA message may represent our best effort to establish on a more visceral level a meaningful connection with extra-terrestrial life.

To complement the efforts of NASA, the METI organization, (Messaging Extra-terrestrial Intelligence), has, since 2017, been broadcasting music to Luyten's Star, a red dwarf in the constellation Canis Minor located approximately 12.36 light-years (3.79 parsecs) from Earth. The signal consists of 33 encoded compositions by various musicians, plus a scientific and mathematical tutorial on how to interpret the messages. The star has a planet in its circumstellar habitable zone. It is one of the closest planets to Earth that may support life as we know it, and so, we could get a response as soon as 2042. Mark your calendars.

In the same year, 2017, EISCAT (the European Incoherent Scatter Scientific Association) broadcast a similar message from Tromso, Norway, to Luyten's Star featuring the first 33 prime numbers, accompanied by music that was based on the algorithmic principles of chemistry and physics. Anyone receiving the message should immediately recognize it as an invitation to join us in an interstellar jam session. If a mixtape is sent back to Earth, it will arrive no sooner than the response to METI. 2042 might yet be a banner year for interstellar communication.

The astronomers who were charged with the responsibility to select and send these communications were motivated to create musical missives that would be calming, comforting, inspiring, and strengthening. These are just the kinds of messages one would want to send to an alien race whose reception and response could not be ascertained beforehand.

To that point, Steven Hawking warned against trying to make contact at all, because of the potential risks to humanity. A distant civilization in another star system might be billions of years more advanced than us, he cautioned, and might view us as inferior, weak, or even insignificant. We might be considered an easy mark if colonial expansion were on their agenda. Hawking said: "They might be vastly more powerful and see us as nothing more than bacteria." He cited the example of Christopher Columbus, observing of the circumstances surrounding his "discoveries" that "things didn't turn out so well" for the native Americans. The potentially bad news for us could be that, for more than a century, we've been indiscriminately broadcasting the electro-magnetic coordinates of our exact location in the cosmos.

So, maybe it would be prudent to turn our communication efforts to music, and hope that the reception of our transmissions will exert a calming influence and intuitively provide evidence of our good intentions. If we reach out to touch our celestial neighbors through the medium of music, perhaps we'll brush up against the face of God at the same time. After all, even before the foundations of the Earth were laid, there was celestial music, when "the morning stars sang together, and all the sons of God shouted for joy." (Job 38:7). 1 Chronicles 16:42 speaks of "trumpets and cymbals … and musical instruments of God." The Psalms were often accompanied by "the musical instruments of David, the man of God." (Nehemiah 12:36). Harmonic strains announced the birth of the Savior. You can almost hear "a multitude of the heavenly host praising God, and saying Glory to God in the highest, and on earth peace, good will toward men." (Luke 2:13-14).

Even during The Dark Ages, in the 9th and 10th centuries, medieval monasteries in the Frankish lands of western and central Europe organized, codified, and notated Gregorian chants, and at the beginning of the Renaissance, madrigals set secular texts to music. Unfortunately, the period between the fall of Rome and feudalism in the 12th century, before the full flowering of the Renaissance, was largely devoid of music in the traditional sense, with the possible exception of the aforementioned chants, and that of troubadours. This underscores the delicate relationship between harmonic music and the flowering of the Spirit.

Between the 12th and the 17th centuries, minstrels in Western Europe wandered the land entertaining with music that later developed into more formal orchestrations. Gothic cathedrals were designed to accommodate choral singing that in the modern age finds expression in music that inspires millions. On the radio since July 15, 1929, the Tabernacle Choir at Temple Square is the longest-running broadcast in the world. It helps us to remember the ancient admonition to "sing unto the Lord." (1 Chronicles 16:23). We "make a joyful noise … and rejoice and sing praise." (Psalms 98:4).

In the dark recesses of memory, we "make sweet melody, (and) sing many songs, that (we may) be remembered." (Isaiah 23:16). We follow the example of the Apostle Paul, who wrote: "I will sing with the spirit, and I will sing with … understanding." (1 Corinthians 14:15). The heavens, the earth, and all that are in them join in the strains. Isaiah urged: "Sing, O heavens; and be joyful, O earth; and break forth into singing, O mountains." (Isaiah 49:13).

Joseph Smith exhorted his brethren: "Let your hearts rejoice and be exceedingly glad. Let the earth break forth into singing. Let the sun, moon, and the morning stars sing together and let all the sons of God shout for joy!" (D&C 128:22-23). Lehi "saw the heavens open, and he thought he saw

God sitting upon his throne, surrounded with numberless concourses of angels in the attitude of singing and praising their God." (1 Nephi 1:8). King Benjamin longed for his "immortal spirit (to) join the choirs above in singing the praises of a just God." (Mosiah 2;28). In metaphor, Alma illustrated our relationship with God, by talking about singing the song of redeeming love. (See Alma 5:26). Mormon looked forward to the time when he could "dwell in the presence of God in his kingdom, to sing ceaseless praises with the choirs above, unto the Father, and unto the Son, and unto the Holy Ghost." (Mormon 7:7).

Sending greetings that have been matched to uplifting music across the vast reaches of the galaxy could have a profoundly positive influence on those who receive the messages. But like the siren song of Greek antiquity, we must remember that men go mad when they hear the tempting strains of unwholesome harmonies. The Adversary has many pen names, but it is chilling, nevertheless, to discover the identity of the author of many of the lyrics set to music that we hear today. He is a master deceiver who knows "the one song everyone would like to learn; the song that is irresistible; the song that forces men to leap overboard in squadrons; the song nobody knows because anyone who has heard it is dead." (Margaret Atwood, "Siren Song," 1976).

Of course, this is the Siren Song of "the nymphs who had the power of charming by their song all who heard them, so that mariners were impelled to cast themselves into the sea to destruction. Circe directed Ulysses to stop the ears of his shipmates with wax, so that they should not hear the strains; to have himself bound to the mast, and to enjoin his people, whatever he might say or do, by no means to release him till they should have passed the Sirens' island. Ulysses obeyed these directions. As they approached, the sea was calm, and over the waters came notes of music so ravishing and attractive that Ulysses struggled to get loose and, by cries and signs to his people, begged to be released; but they, obedient to his previous orders, sprang forward and bound him still faster. They held to their course, and the music grew fainter, till it ceased to be heard, when with joy Ulysses gave his companions the signal to unseal their ears; and they relieved him from his bonds." (Homer, "Ulysses").

Subtle influences such as the Sirens' Song are at work to compromise the high standard set by faith-promoting music. We live in a vulgar and profane world that is demeaning to character. In too many ways, we secularize sacred things, and trivialize the celestial. We denigrate it with careless words and with thoughtless musical expressions.

There are at least 37 different ways the hymns of the Lord's Church guard against "trivializing the celestial." Each selection in the hymnal is accompanied by a suggestion regarding the emotions with which it should be sung. Each sentiment ennobles the selection with an inspiring quality that invites the Spirit in a particular way. These are just the qualities of music that could be perfectly interwoven within an interstellar invitation to anyone who might be listening, to share the harmony of the heavens.

<u>Boldly</u>. We sing with boldness when we are confident. The book of Acts records that when the Apostles "had prayed, the place was shaken where they were assembled together; and they were all filled with the Holy Ghost, and they spake the word of God with boldness." (Acts 4:31). Brightly.

We sing brightly when we are animated. Isaiah prophesied that the day would come when the Gentiles would come to the Light of Christ and the leaders of government to the brightness of his rising. (See Isaiah 60:3).

Calmly. We sing calmly when we are sure of ourselves. Of the Savior, the Psalmist wrote: "He maketh the storm a calm, so that the waves thereof are still." (Psalms 107:29).

Cheerfully. We sing cheerfully because we do not allow circumstances to determine our emotions, and when we are at peace, even though we may be surrounded by turmoil. The Savior reassured His disciples: "In the world, ye shall have tribulation, but be of good cheer. I have overcome the world." (John 16:33).

Confidently. We sing confidently because we have a positive self-image. "Fiddler on The Roof's" wise old Tevya told his daughters: "In Anatevka, God knows who you are, and what he expects you to become." (Joseph Stein). As Proverbs teaches: "The Lord shall be thy confidence." (Proverbs 3:26).

With Contemplation. We pause in our busy lives to ponder the solemnities of eternity, and to "contemplate the word of the Lord." (D&C 124:23).

With Conviction. We sing with conviction when we are certain of our course. Sometimes, we are "convicted by (our) own conscience." (John 8:9). When this happens, we are resolute and are sure of our direction, as our faith leads us to purposeful repentance, forgiveness, cleanliness, and holiness.

With Devotion. We sing with devotion because we have consecrated our hearts, might, mind, and strength to the Lord. All He requires is that we "devote all (our) service in Zion; and in this, (we) shalt have strength." (D&C 24:7).

With Dignity. We know who we are, and we intuitively sense "the excellency of dignity." (Genesis 49:3).

Earnestly. We are blessed with the power to be true to our covenants, remembering that the Lord has enjoined us to earnestly seek the best spiritual gifts. (See D&C 46:8).

With Emotion. As we realize that God is "ready to pardon, gracious and merciful, slow to anger, and of great kindness," we are overcome with emotion. (Nehemiah 9:17).

Energetically. When we are anxious to be about our Father's business, we channel our energy and focus it on the tasks at hand. As Alma exhorted his son, so are we similarly encouraged. We "turn to the Lord with all (our) mind, might, and strength." (Alma 39:13).

With Energy. We are quickened by the Spirit and it gives us voice. We are as Nephi, who exhorted his brethren "with all the energies of (his) soul, and with all the faculty which (he) possessed." (1 Nephi 15:25).

Enthusiastically. We are caught up in eternal burnings when God shows us "His glory and his greatness, and we (hear) his voice out of the midst of the fire." (Deuteronomy 5:24).

Expressively. We sing with expression because we are finely attuned to the whisperings of the Spirit and respond appropriately. The Savior Himself was the personification of harmony with His Father, "and the express image of his person," when He "sat down on the right hand of the Majesty on high" as a resurrected and perfected being. (Hebrews 1:3).

Exultantly. When we are carried away with the conviction that the Savior will overcome the world, we "rejoice and exult in the hope, and even know, according to the promises of the Lord, that (we will be) raised to dwell at the right hand of God, in a state of never-ending happiness." (Alma 28:12).

Fervently. When we are zealous in our discipleship, we are as the Apostles of old, who taught a disciple who "was instructed in the way of the Lord; and being fervent in the spirit, he spake and taught diligently the things of the Lord." (Acts 18:25).

Firmly. When we are quietly confident that our course leads back to our Heavenly Father, we are as the Sons of Helaman, of whom it was written: "Their minds (were) firm, and they (did) put their trust in God continually." (Alma 57:27).

Gently. When we are hesitant to disrupt the sweetness of the Spirit that we feel, we are as the Psalmist, who exclaimed: "Thou hast also given me the shield of thy salvation: and thy right hand hath holden me up, and thy gentleness hath made me great." (Psalms 18:35).

Humbly. We sing with humility because the sweet strains break our hearts with contrition. "If my people, which are called by my name, shall humble themselves, and pray, and seek my face, and turn from their wicked ways," said the Lord, "then will I hear from heaven." (2 Chronicles 7:14).

Joyfully. When our desire to praise God from the rooftops is unrestrained, we "make a joyful noise unto the Lord," and we "make a loud noise, and rejoice, and sing praise." (Psalms 98:4).

Jubilantly. We sing jubilantly when it dawns upon us that as many as call upon God will be saved by His matchless grace. We are as the Israelites of old who were instructed: "Then shall ye cause the trumpet of the jubilee to sound ... throughout all your land. And ye shall ... proclaim liberty throughout all the land unto all the inhabitants thereof."
(Leviticus 25:9-10).

Lightly. The Lord will carry our burdens for us, and "our light affliction, which is but for a moment, worketh for us a far more exceeding and eternal weight of glory." (2 Corinthians 4:17).

Majestically. When we realize that we are sons and daughters of a noble birthright, everything snaps into sharp focus. The children of God "speak of the glory of (His) kingdom and talk of (His) power; to make known to the sons of men His mighty acts, and the glorious majesty of His kingdom." (Psalms 145:11-12).

<u>Meekly</u>. Our tender feelings reflect the nature of God and "the meekness and gentleness of Christ." (2 Corinthians 10:1). The Savior Himself said: "Take my yoke upon you and learn of me; for I am meek and lowly in heart: and ye shall find rest unto your souls." (Matthew 11:29).

<u>Peacefully</u>. The sanctuary of music is a refuge from the cares of the world, and when we "publish peace," these "tidings of good … declare unto the people that the Lord reigneth." (Mosiah 27:37).

<u>Prayerfully</u>. Melodious refrains find their way across the heavens to reach God's listening ear. "The song of the righteous," He said, "is a prayer unto me." (D&C 25:12).

<u>Reflectively</u>. Our hymns are clothed with unprecedented power as we contemplate our place in the cosmos, "reflecting upon the great atoning sacrifice that was made by the Son of God, for the redemption of the world." (D&C 138:2).

<u>Resolutely</u>. With the righteous, we resolve to serve God in every circumstance. Joseph Smith declared to his brethren: "I receive you to fellowship, in a determination that is fixed, immovable, and unchangeable, to be your friend and brother through the grace of God in the bonds of love, to walk in all the commandments of God blameless, in thanksgiving, forever and ever." (D&C 88:133).

<u>Smoothly</u>. We sing smoothly when it dawns on us that no wind can blow except to fill our sails. We are borne up as upon the wings of eagles. (See D&C 124:18). When we sing about Gospel principles, we cannot feel the bumps that lie before us on the highways and byways of life. Our music is seamlessly integrated with principled behavior and quietly validates our commitment. As we run life's race, there are no false starts, and there is no dropping of the baton as it is passed on. Symmetry gives us the quiet confidence to smoothly endure to the end in righteousness.

<u>Solemnly</u>. As we approach God and tread on sacred ground, we treasure His knowledge "up in (our) hearts and let the solemnities of eternity rest upon (our) minds." (D&C 43:34).

<u>With Spirit</u>. Christ is the way, the truth, the life, and the light, and without Him, our world that sits far out on the Orion Spur of the Milky Way Galaxy would be dreary, tedious, and monotonous. "The Spirit of God burns like a fire, even as "the latter-day glory begins to come forth. The visions and blessings of old are returning, and angels are coming to visit the earth." (William W. Phelps, Hymn #2, "The Spirit of God").

<u>Tenderly</u>. When we think of the Savior's sacrifice that was motivated by His love for us, we plead, as did the Psalmist: "Withhold not thou thy tender mercies from (us), O Lord; let thy loving kindness and thy truth continually preserve (us)." (Psalms 40:11)

<u>Thankfully</u>. Every time He gives us voice, we are as the Israelites of old who "sang together … in praising and giving thanks unto the Lord, because he is good." (Ezra 3:11). David exhorted his people: "Rejoice in the Lord, ye righteous; and give thanks at the remembrance of his holiness." (Psalms 97:12).

<u>Thoughtfully</u>. As "weary travelers (we) find health and safety while (we) contemplate the word of the Lord." (D&C 124S:23). The Savior directed Joseph Smith to establish the Nauvoo House as "a delightful habitation for man, and a resting-place for the weary traveler, that he may contemplate the glory of Zion." (D&C 124:60).

<u>Triumphantly</u>. We recall our sustained efforts, confident that we have fought a good fight, have run our race, and have done our best. We remember Moroni's farewell: "I soon go to rest in the paradise of God, until my spirit and body shall again reunite, and I am brought forth triumphant through the air, to meet you before the pleasing bar of the great Jehovah." (Moroni 10:34).

<u>Vigorously</u>. The Spirit quickens and energizes us to be up and about our Father's business. After Mormon had exhorted his brethren, he was pleased to report that his "words did arouse them somewhat to vigor." (Mormon 2:24).

<u>Worshipfully</u>. When we are truly aware of our surroundings, we realize that we stand in holy places. On Sinai, Moses was told: "Put off thy shoes from off thy feet, for the place whereon thou standest is holy ground." (Exodus 3:5). The Wise Men of the East were inspired to inquire of Herod: "Where is he that is born King of the Jews? For we have seen his star in the east and are come to worship him." (Matthew 2:2). Our homes invite us to "stand in holy places and be not moved." (D&C 45:32).

Heavenly Father and Jesus Christ love music. The Savior revealed how deeply it touched Him, when He said: "My soul delighteth in the song of the heart." (D&C 25:12). Its inherent power expands our focus outward, that we might embrace others on the stage of shared experience. Wholesome music tugs at our heartstrings, activating resounding harmonies that stir our souls and reach all the way to the heavens.

Certainly, both the cosmos and celestial worlds are filled with music. As Helen Keller wrote: "I believe that my home there will be beautiful with colour, music, and speech of flowers and faces I love." (Midstream). One day in the not-too-distant future, as they gather from galaxies both near and far away, the children of God will be reunited at melodious family reunions. The atmosphere will resound with a musical language that enjoys universal application, appreciation, expression, recognition, and understanding.

When, in just 24 days, George Frideric Handel created the 259 pages of musical score that comprise "The Messiah," the notes came to him so quickly that he could barely keep up, as he furiously scratched out the oratorio on whatever paper was handy. After he had written the "Hallelujah Chorus" in a fervor of divine inspiration, he exclaimed that he had "seen all heaven before him." On the last page of the manuscript, in acknowledgement of his own puny efforts, he wrote the letters "SDG" that stood for "Soli Deo Gloria" or "To God alone the glory."

On a much smaller scale, many of us have had similar experiences with light and knowledge as they relate to music. We have felt the Gods smiling upon us as we have been permitted, through music, to catch a glimpse of the flurry of activity that takes place just beyond the parted veil. In our feeble attempts to Dance with The Stars, it would be wonderful if nurturing and stimulating revelatory experiences that were accompanied by music could find their way into our extra-terrestrial expressions.

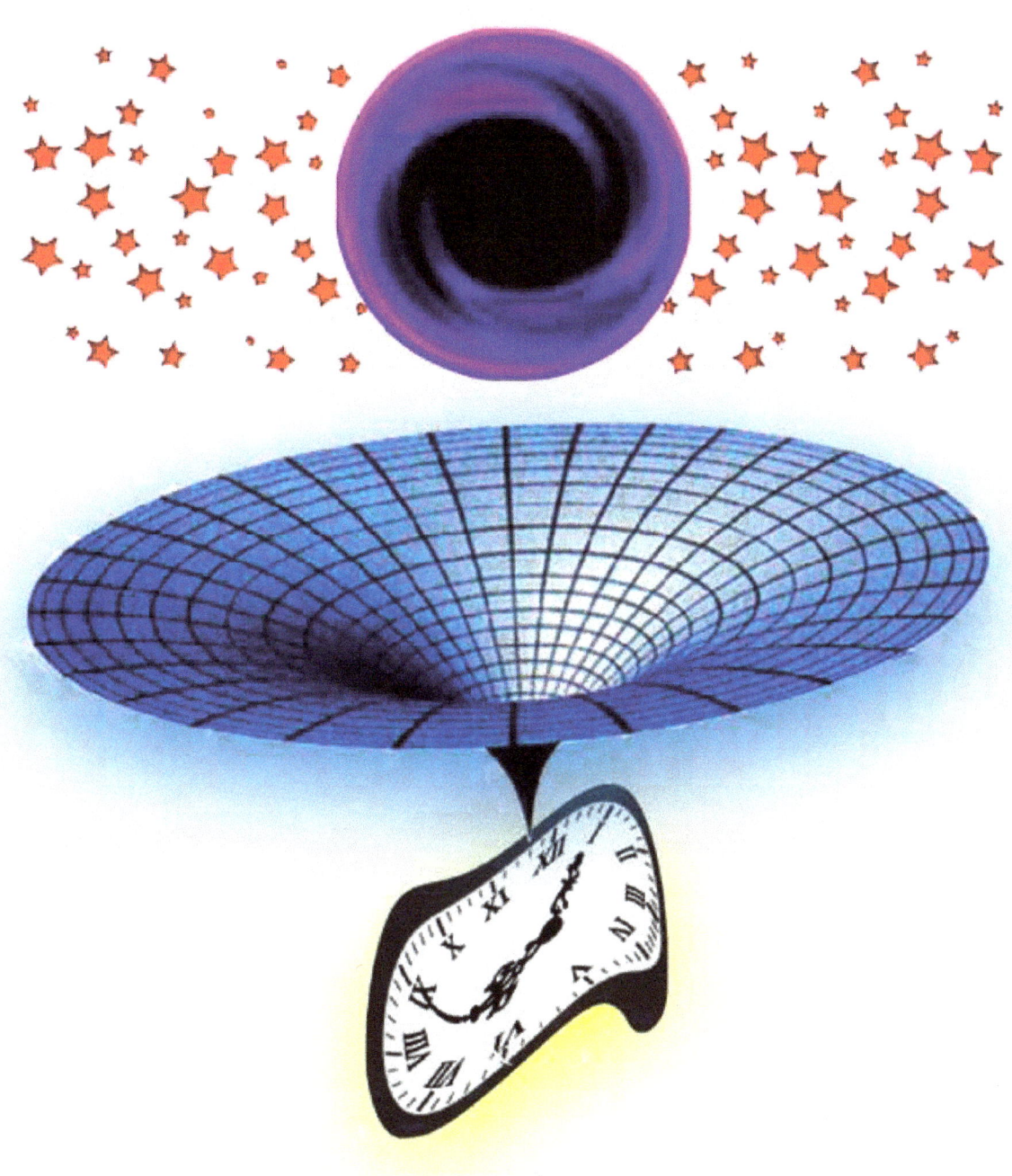

Chapter Eight

Habla Usted Inglés?

We have no idea what form the language of extra-terrestrial beings might take; the only real clues we have come from own religious experience and history. We know that, in the beginning, "the whole earth was of one language, and of one speech." (Genesis 11:1). But, after the Tower of Babel was constructed to circumvent God as a profane ziggurat designed to reach all the way to heaven, He said: "Let us go down, and there confound their language, that they might not understand one another's speech, (and) the Lord did there confound the language of all the earth: and from thence did the Lord scatter them abroad upon the face of all the earth" (Genesis 11:7 & 9).

As a result, the inhabitants of the Earth could not understand each other, but in time, their speech coalesced into what are today roughly 6,500 distinct languages. Sadly, in our day, voice communication frequently breaks down, and intuitive, insightful, inspirational, and revelatory nonverbal communication with the Spirit seems to be even more tenuous. As Laman and Lemuel revealingly complained long ago: "Behold, we cannot understand the words which our father hath spoken." (1 Nephi 15:7). Later, Paul explicitly taught the Corinthian Saints that "the natural man receiveth not the things of the Spirit of God; for they are foolishness unto him: neither can he know them, because they are spiritually discerned." (1 Corinthians 2:14).

If we better understand the different categories of higher-level verbal and non-verbal correspondence between ourselves and heavenly beings, we might be able to extrapolate potential avenues of communication between ourselves and corporeal extra-terrestrials and, thus, be better prepared for First Contact.

<u>The Adamic Language</u>

When the Lord comes again, He will restore the Adamic tongue: "For then will I turn to the people a pure language, that they may all call upon the name of the Lord to serve him with one consent." (Zephaniah 3:8-9). According to Jewish tradition, the Adamic language was spoken by our first parents in the Garden of Eden. (See Genesis 2:19). It is believed by many to be the language that God used to converse with them at that time.

Joseph Smith declared the Adamic tongue to have been "pure and undefiled." (Moses 6:6). The interpretation of this verse in the Pearl of Great Price leads many to believe it to be the language of heaven, as well. (See "Adamic Language" in "Encyclopedia of Mormonism" p. 18-19). Words thought to be derived from the Adamic language include "deseret" (or honeybee - see Ether 2:3), and "Ahman" (or God - see D&C 78:20). In addition to these two scriptural allusions to the Adamic language, there are many additional examples of beings from the unseen world speaking to mortals. A sampling includes the following:

<u>Direct experience with God the Father</u>

On the Mount of Transfiguration, "there was a cloud that overshadowed" Jesus, Peter, James, and John, "and a voice came out of the cloud, saying, This is my beloved Son: hear him." (Mark 9:7).

At the baptism of Jesus, "the Holy Ghost descended in a bodily shape like a dove upon him, and a voice came from heaven, which said, Thou art my beloved Son; in thee I am well pleased." (Luke 3:22).

<u>Direct experience with God the Father and His Son Jesus Christ</u>

Of his Sacred Grove experience, Joseph Smith recalled: "I saw a pillar of light exactly over my head, above the brightness of the sun, which descended gradually ... When the light rested upon me, I saw two Personages, whose brightness and glory defy all description, standing above me in the air. One of them spake unto me, calling me by name and said, pointing to the other - This is My Beloved Son, Hear Him!" (Joseph Smith History 1:16-17).

Rather than unnecessarily complicating his account of the First Vision by including a primer on spiritual forms of communication, Joseph Smith might have consciously omitted any reference to the language used by Heavenly Father in the Sacred Grove. It is entirely possible that they communicated, not in English, but rather in the pure and undefiled language of heaven. (See 1 Nephi 15:7 and 1 Corinthians 2:14, cited above). Joseph could also have simply been unaware of the language of the First Vision. As he later described the revelatory process, "by the power of the Spirit (his) eyes were opened and (his) understandings were enlightened so as to see and understand the things of God." (D&C 76:12). Note also that the Apostle Paul made a point, when describing to Agrippa his "Road to Damascus" experience, that the Savior had communicated with him "in the Hebrew tongue", as if to differentiate it from other possible languages. (Acts 26:14). Elsewhere in the Book of Acts, Luke made it clear that those who were with Saul at the time did not hear anything at all. They only saw "the light, and were afraid, but they heard not the voice of him that spake to (Saul)." (Acts 22:9).

<u>Direct experience with Jehovah</u>

"God called unto (Moses) out of the midst of the bush, and said, Moses, Moses ... Draw not nigh hither: put off thy shoes from off thy feet, for the place whereon thou standest is holy ground ... I am (Jehovah) the God of thy father, the God of Abraham, the God of Isaac, and the God of Jacob." (Exodus 3:4-6).

<u>Direct experience with Jesus Christ</u>

"And no tongue can speak, neither can there be written by any man, neither can the hearts of men conceive so great and marvelous things as we both saw and heard Jesus speak; and no one can conceive of the joy which filled our souls at the time we heard him pray for us unto the Father." (3 Nephi 17:17).

Manifestations of the Spirit

Joseph Fielding Smith said that the Holy Ghost leaves an indelible impression upon our souls that is not easily erased. "It is Spirit speaking to spirit, and it comes with convincing force. A manifestation of an angel, or even of the Son of God himself, would impress the eye and mind, and eventually become dimmed, but the impressions of the Holy Ghost sink deeper into the soul and are more difficult to erase." ("Answers to Gospel Questions," 2:151).

With reverential awe, Hugh B. Brown recalled similar manifestations of the Spirit. "Sometimes during solitude," he declared, "I hear truth spoken with clarity and freshness; uncolored and untranslated, it speaks from within myself in a language original but inarticulate, heard only with the soul."

A Still Small Voice

And an angel said: "Go forth and stand upon the mount before the Lord. And behold, the Lord passed by, and a great and strong wind rent the mountains, and brake in pieces the rocks before the Lord; but the Lord was not in the wind: and after the wind an earthquake; but the Lord was not in the earthquake: And after the earthquake a fire; but the Lord was not in the fire: and after the fire a still small voice." (1 Kings 19:11-12, see 1 Nephi 17:45, & D&C 85:6).

Angels

In his History, Joseph Smith described his first encounter with an angel who had come from the presence of God: "I discovered a light appearing in my room, which continued to increase until the room was lighter than at noonday, when immediately a personage appeared at my bedside, standing in the air, for his feet did not touch the floor. He had on a loose robe of most exquisite whiteness. It was a whiteness beyond anything earthly I had ever seen; nor do I believe that any earthly thing could be made to appear so exceedingly white and brilliant. His hands were naked, and his arms also, a little above the wrist; so, also, were his feet naked, as were his legs, a little above the ankles. His head and neck were also bare. I could discover that he had no other clothing on but this robe, as it was open, so that I could see into his bosom. Not only was his robe exceedingly white, but his whole person was glorious beyond description, and his countenance truly like lightning. The room was exceedingly light, but not so very bright as immediately around his person. When I first looked upon him, I was afraid; but the fear soon left me. He called me by name and said unto me that he was a messenger sent from the presence of God to me, and that his name was Moroni." (J.S.H. 1:30-33).

Metaphysical manifestations of seraphim, cherubim, thunderings, lightnings, earthquakes, and music.

"Then flew one of the seraphims unto me, having a live coal in his hand, which he had taken with the tongs from off the altar: And he laid it upon my mouth, and said, Lo, this hath touched thy lips." (Isaiah 6:6-7)

"I placed at the east of the Garden of Eden, cherubim and a flaming sword, which turned every way to keep the way of the tree of life." (Moses 4:31).

"And it came to pass ... that there were thunders and lightnings, and a thick cloud upon the mount, and the voice of the trumpet exceeding loud; so that all the people that was in the camp trembled. And mount Sinai was altogether on a smoke, because the Lord descended upon it in fire: and the smoke thereof ascended as the smoke of a furnace, and the whole mount quaked greatly." (Exodus 19:16-18). "And all the people saw the thunderings, and the lightnings, and the noise of the trumpet, and the mountain smoking: and ... they removed and stood afar off." (Exodus 20:18).

"I was in the Spirit on the Lord's day, and heard behind me a great voice, as of a trumpet." (Revelation 1:10). "The first voice which I heard was as it were of a trumpet talking with me, which said, Come up hither, and I will shew thee things which must be hereafter." (Revelation 4:1). It stirs the blood to think of the voices of trumpets speaking to us in a musical language that is inarticulate and indescribable, and yet is irrefutable.

"And I heard a voice from heaven, as the voice of many waters, and as the voice of a great thunder: and I heard the voice of harpers." (Revelation 14:2).

In The Book of Mormon, especially in 3 Nephi in those chapters that concern the ministry of Jesus Christ among the Nephites:

"After this manner doth the Lord God work among the children of men. For (He) giveth light unto the understanding; for he speaketh unto men according to their language." (2 Nephi 31:3). "I perceive that ye ... cannot understand all my words which I am commanded of the Father to speak unto you at this time. Therefore, go ye unto your homes, and ponder upon the things which I have said, and ask of the Father, in my name, that ye may understand, and prepare your minds for the morrow, and I come unto you again." (3 Nephi 17:2-3).

"Behold, he prayed unto the Father, and the things which he prayed cannot be written, and the multitude did bear record who heard him." (3 Nephi 17:15). "And the multitude did hear and do bear record; and their hearts were open, and they did understand in their hearts the words which he prayed. And tongue cannot speak the words which he prayed, neither can be written by man the words which he prayed." (3 Nephi 19:32). "So great and marvelous were the words which he prayed that they cannot be written, neither can they be uttered by man." (3 Nephi 19:33-34).

<u>Forms of communication between ourselves and mortal beings from extra-terrestrial worlds</u>.

Of course, we have no idea what that communication might sound like, or for that matter, what it might look like, or feel like, or what flavor it might be, or what aroma it might convey, should humanity one day make First Contact with beings like (or unlike) ourselves, from elsewhere in the galaxy. That interaction could be auditory, or it might just as plausibly be linked to our other somatic senses of sight, touch, taste, or smell, or it could be linked to the expression of more subtle sensations and perceptions.

For example, in our everyday world, expressions of happiness sound like children laughing in the park, look like a bright smile, make us feel warm and cozy inside, taste like an ice cream sundae, and smell like bread baking in the oven. Apples sound crunchy when biting into them, look like big red balls, feel smooth and hard, taste sweet and juicy, and smell like cider. These familiar senses convey meaning and are all equally powerful forms of communication.

Sound, sight, touch, smell, and taste are the palpable passport stamps that attest to our interaction with the world around us, and especially to the different ways that we can have meaningful relationships with others. Our brains are powerful blenders that take these sensations and whip them up into frothy virgin piña coladas of perception that become our own inimitable windows on the world. They are our own personalized concoctions, for their recipes are proprietary. Life experiences create zesty signature specialty drinks complete with little umbrellas to protect us from rain showers. They are topped with mouth-watering maraschino cherries to hold our interest, and whipped cream on top keeps us coming back to the Server asking for more, please! If you are now salivating, that is simply a confirmation of the ability of our somatic senses (in this case, sight) to trigger perception within our brains and elicit dramatic physiological reactions.

But, as common as our five physical senses are to all of us, we communicate with others in subtly different ways. We were never meant to be cookie-cutter copies of each other. Closely associated with innovation and experimentation in the learning laboratory of life, our own personal experiences testify to the recognition by our Creator of our individuality. It is within His divinely conceived and ordained process of the marriage of sense and perception that our personalities enjoy an elasticity that adds a delightful element of uncertainty to the conversation. No-one can ever know exactly where it will be headed. The resulting originality allows our dialogue to capitalize on the gentle tug of non-conformity. Just as wind and water shape the landscape, the Master Potter fashions our mortal clay with the needle tool of experience, the potter's rib of participation, the wire cutting tool of practice, and the trimming loops and sponges of performance. The result is stimulating creativity and diversity among populations of similarly minded creatures.

As they are fired in the cauldron of experience, the pieces of the puzzle, that in their sum and substance define us, express their unique characters. We are works in progress, and not all the chapters have been written; we are stanzas in an unfinished symphony and are continually evolving in an eternal progression. That is one of the things that makes our anticipation of First Contact so intriguing, refreshing, and exciting.

We understand that Shakespeare must have been expressing something more than just intellectual appreciation or acknowledgement, when he wrote: "What a piece of work is a man! How noble in reason, how infinite in faculty! In form and moving how express and admirable! In action how like an angel, in apprehension how like a god! The beauty of the world. The paragon of animals." ("Hamlet", Act 2, Scene 3).

Truly, in God, "we live and move, and have our being" through sense and perception that are blended into a refreshing elixir that prevents us from becoming too comfortable with the stale or flat beverages that we sometimes seek for refreshment on the highways and byways of life. (Acts 17:28). Sense and perception form an unlikely union, that by Intelligent Design has been created to upset the status-quo, enlarge our experience, weather every storm, and meet every challenge. At the end of the day, it is "by the experiment of this ministration" utilizing all of our senses, that we are able to "glorify God for (our) professed subjection unto the gospel of Christ." (2 Corinthians 9:13).

It was ordained in the heavens that, even as we internalize the doctrine of individuality, we feel the exertion of an equalizing influence, for, ultimately, throughout the cosmos, "all are alike unto God." (2 Nephi 26:33). But at the same time, sense and perception enhance the painstakingly crafted formula of a distinctive bouquet that make each of us unique. When we do establish First Contact, it is going to be a mind-expanding experience of unprecedented proportion, because extra-terrestrials are sure to flavor the celebration with their own individuality and unique qualities.

We might find that some of our alien guests will prefer the libation of red wines that tend to have a rich and bitter taste. Others might request the light and sweet taste of white wine. But we cannot pre-determine if the communication that will accompany First Contact is going to come via the sense of taste, or from any of the other somatic senses, for that matter. There are myriad ways of interacting with others. These include thermoception (temperature), proprioception (body position), nociception (pain), equilibrioception (pain), and chronoception (time). Together, like hormones, they subtly catalyze our interactive awareness. To add even more distinctive flavor to the stew that is simmering on the cauldron of communication, there are other senses that we do not have, or have only in diminished capacity. These include, but are not limited to, the ability to detect electrical and magnetic fields, polarized and infrared light, water pressure, pheromones, and the ability to utilize sonar.

At the end of the day, we do not have a lock on the senses that we enjoy. They are not uniquely ours. Bears have a better sense of smell, eagles a better sense of sight, catfish (believe it or not) a better sense of touch and taste, and dogs a better sense of sound. But those we do have seem to be harmoniously in perfect balance and are individually tailored to our specific needs. Only time will tell whether extra-terrestrial guests on the interstellar stage will exhibit unusual senses that lie somewhere beyond our every-day experience. If they do, they will surely introduce to the conversation extraordinary new perspectives, and they might even become the life of the party. There is unlikely to be a dull moment, or a lull in the conversation, during the festivities.

It could be argued that our central nervous systems define us, and that all the rest is just bubble-wrap packaging. Breaking it all down (or popping the bubble-wrap), roughly 30 percent of the neurons in our cerebral cortex are devoted to vision, 8 percent to touch, 3 percent to hearing, and 0.1% to smell. It is unknown how many of the brain's neurons in the gustatory cortex are related to taste, but if our enjoyment of milk chocolate is any indication, it must be a lot.

Roughly 100 billion neurons make up our cortical grey matter, with 100 trillion connections (synapses) and a processing capacity of ten trillion instructions per second. The electrical output that would be required to simulate our brain function in a laboratory setting is equivalent to ten megawatts. A phenomenal amount of activity is taking place between our ears to accommodate virtually limitless perceptions that take the philosophical supposition "cogito ergo sum" to incomprehensible levels.

Our senses are real, and when our bodies are in homeostasis, or in a state of balance or equilibrium, we experience all of them, generally unconsciously (since we can only focus our attention on one thing at a time). They blend in harmony to give us all the information we need to interact with our environment, and with terrestrial sentient carbon-based life forms.

But they do more than just that. Within our brains, there is a magical transformation of raw data into perception, and in the process, our lives are enriched. Perception confirms that the sum is greater than its individual parts. It sweeps aside the simple math of stimulus and response, nullifying the argument that we merely react to our surroundings, and it introduces the intriguing element of uncertainty that gives vitality and vibrancy to life, and new meaning to self-awareness. If we stretch our minds to galactic proportions, perception opens new possibilities relating to interactions with our interstellar eighbors. Perception opens our minds to unknown possibilities of existence that we may have never considered, but that one day may be of unprecedented importance.

Thanks to our culture's fascination with science fiction in motion pictures, we have interactive dramatizations relating to how we might communicate with extra-terrestrials. An example that comes to mind is the language of the Klingon Empire, in the Star Trek Universe. Klingon is a spoken language, albeit one that hurts our ears. When it was created by science fiction aficionados, it was meant to sound alien. Its sentence structure is object-subject-verb, which is the least common construction among all 6,500 actual spoken human languages. In the words of a noted linguist: "Klingon falls just short of being unpronounceable, unlearnable, and barely intelligible." Nevertheless, it is the standard bearer of alien spoken languages, making the fictional Star Trek Universe seem more real. One thing is for certain: If we encounter Klingons during First Contact, Universal Translator technology will sorely be put to the test.

As a footnote, it may never become necessary to learn Klingon in the real world, but just in case, here is a sampling of common phrases: Heghlu'meH QaQ jajvam! (Today is a good day to die!) Eghlu'DI' mobbe'lu'chugh QaQpu' hegh wanI' (Death is an experience best shared.) 'Oy' DaSIQjaj! (May you endure the pain!) Nuq daq yuj da'pol? (Where's the chocolate?) QuSDaq ba'lu''a'? (Is this seat taken?) vIjatlh! (Speak!) Pe'vIl mu'qaDmey! (Curse well!) NuqDaq 'oH puchpa''e'? (Where's the bathroom?) NuqDaq 'oH tach'e'? (Where's the bar?) Tera'ngan Soj lujab'a'? (Do they serve earth food here?) Qut na' HInob. (Give me the salty crystals.) Gagh Sopbe'. (He doesn't eat gagh.) HIja / ghobe' (Yes / No). Dochvetlh vIneH! (I want that!) Hab SoSlI' Quch! (Your mother has a smooth forehead!) Nuqjatlh? (What did you say?) Jagh yIbuStaH! (Concentrate on the enemy!) QaStaH nuq jay'? (What the #$*@ is happening?) Wo' batlhvaD! (For the honor of the Empire!) Tlhingan maH! (We are Klingon!) Qapla'! (Success!)

Chapter Nine

Heptapod Logograms
and The Finger of the Lord

"And it came to pass that ... the Lord stretched forth his hand ... and the veil was taken from off the eyes of the brother of Jared, and he saw the finger of the Lord; and it was as the finger of a man, like unto flesh and blood; and the brother of Jared fell down before the Lord, for he was struck with fear (and) the Lord could not withhold anything from his sight; wherefore he showed him all things, for he could no longer be kept without the veil." (Ether 3:6 & 12:21).

Among many types of terrestrial communication, there are +/- 6,500 spoken languages and +/- 4,000 written languages, not to mention experiential forms of communication such as body language, gestures, eye contact and movement, facial expressions, touch, voice intonation, laughing, crying, and personal space, to name just a few. Among many written forms of communication, logograms are characters that represents a word or a morpheme (the smallest meaningful lexical item in such languages). In the ancient world, cultures that have used logograms include the Egyptians, Mayans, Sumerians, and Akkadians. In the modern world, there are over 50,000 Mandarin logograms, although educated Chinese typically only utilize around 8,000 of those characters.

The New Testament preserves an account when "Jesus stooped down, and with his finger wrote on the ground." (John 8:6). This is the only instance in the scriptures where the Savior wrote something, but there is no supporting documentation relating to what it might have symbolized. In The Book of Mormon, there are two instances where He used His finger to communicate with His disciples. In the first case, "the Lord stretched forth his hand and touched (twelve) stones one by one with his finger." (Ether 3:6). Secondly, after He had taught the people in Zarahemla, "he touched every one of them with his finger." (3 Nephi 28:12). Whether these are examples of experiential language or of written language, either in logograms or in some other form, is unknown, but they have fascinated both linguists and religious scholars for millennia. Linguistics is the discipline that seeks to understand how language works. In the motion picture "Arrival", a linguist named Louise Banks was recruited by government officials who were concerned about the motives of the crew of an alien spaceship that had recently appeared, and then hovered, over a field in rural Montana.

Banks looked for patterns in logograms that were created by the heptapod inhabitants of the ship, and she went about her task by starting with the basics, and then logically working her way up to questions like "What is your purpose on Earth?" In time, she discovered that the aliens had a written language that was circular. Every sentence was conceived all at once, reflecting their non-linear perception of the passage of time.

In the heptapod language, each circular logogram was divided into 12 equal segments. Each segment had a shape that conveyed meaning. The thickness of a line denoted inflection. A hook shape posed a question. Understanding the language, as it turned out, was related to seeing both the present and the future at once.

The creators of "Arrival" took some artistic license with the script, but it is intriguing to consider the possibility of First Contact taking place within the parameters of a similar scenario. For example, in the motion picture "E.T. – The Extra-Terrestrial", each time E.T. extended his finger, it conveyed meaning, and amazing things happened.

As an aside, Stephen Spielberg, who directed "E.T. – the Extra-Terrestrial", was invited by President Ronald Reagan to give a private screening at the White House, with senior government officials in attendance. At the end, as Spielberg remembered it: "The President just stood up and he looked around the room, almost like he was doing a headcount, and he said, "I want to thank you for bringing E.T. to the White House. We really enjoyed your movie," and then he looked around the room again, and said: There are a number of people in this room who know that everything on that screen was absolutely true. He said it without smiling, but the whole room laughed, because he presented it like a joke."

"I don't think he let something slip there," Spielberg recalled. "I think he delivered a joke, but without a twinkle in his eye. I think the joke landed because everybody laughed, but because I'm a little bit of a UFOlogist, I was hoping that there was something more to the joke than met my eye." (However, we'll never know for sure, at least until the government releases the details of First Contact.)

The heptapods in "Arrival" extended their fingers as did both E.T. and the Savior, but they id it to write logograms. They included the following, which formed the basis of lasting communication with their human counterparts. In the motion picture, at least, these logograms changed the course of history, both domestically and galactically, but you'll need to watch the film to find out how that played out.

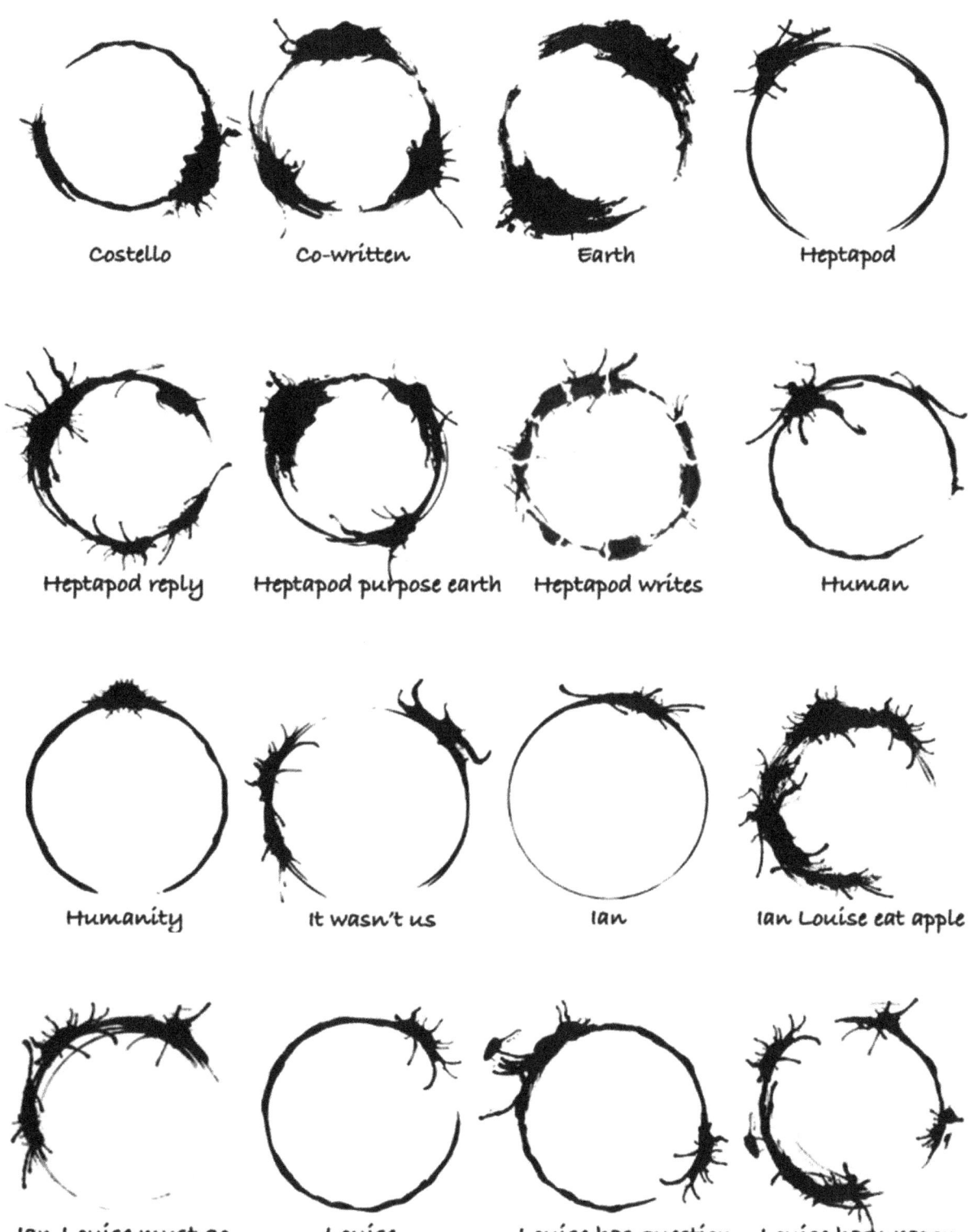

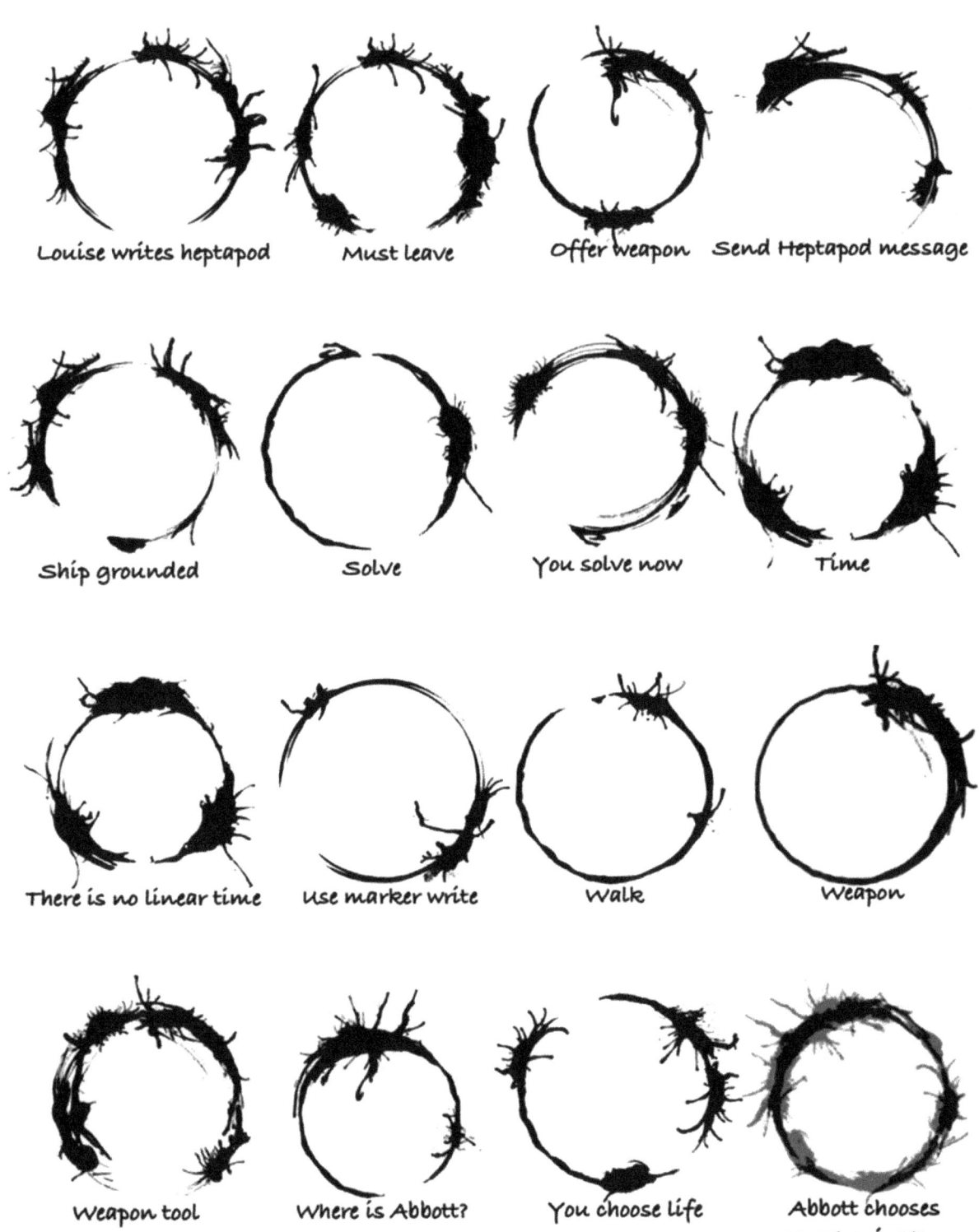

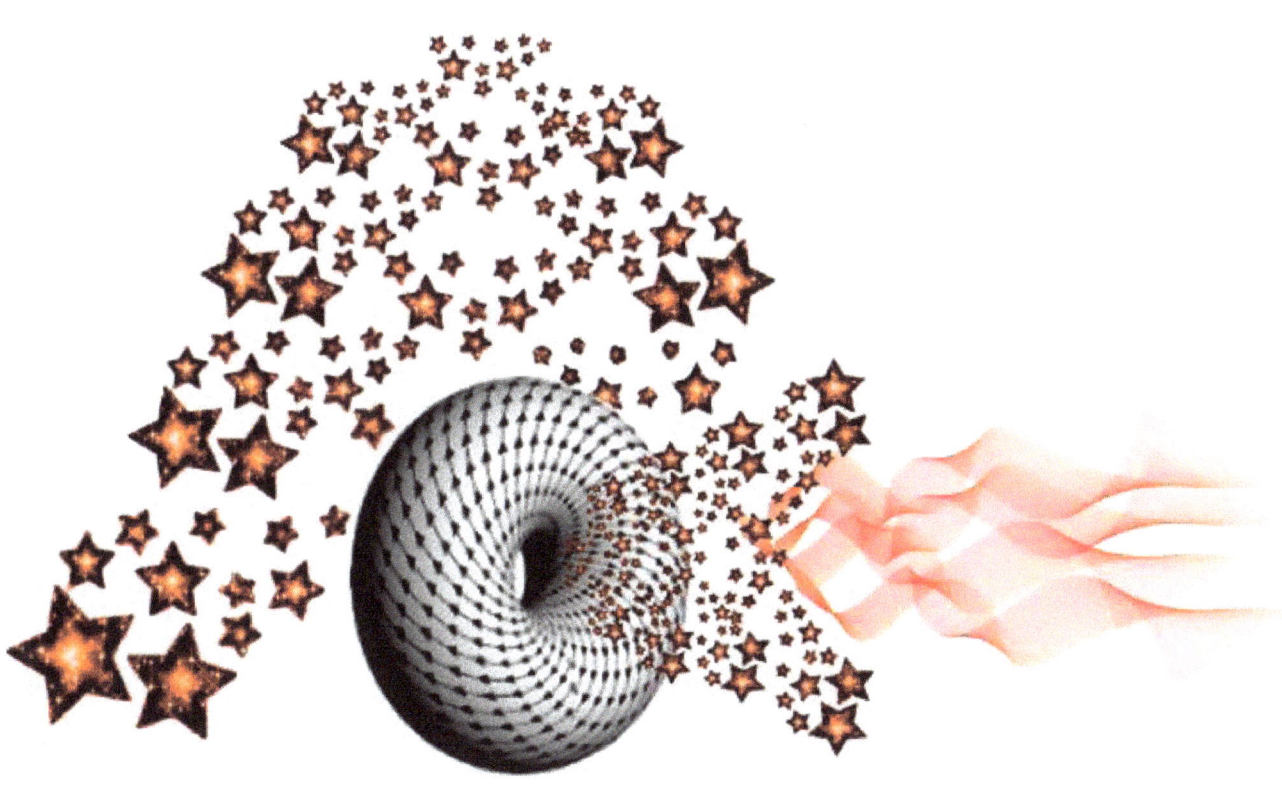

Chapter Ten

Synaesthesia

In our quest to discover what is real, we remember that Joseph Smith, as a prism of the Lord, saw through the clarifying lens of eternity. The light that illuminated his mind and spirit could be broken down into its individual elements of principles and doctrines, ordinances and covenants, and commandments and blessings, as if by a polyhedron with two polygonal faces lying in parallel planes and with the other faces oriented as parallelograms (a prism).

The light of the Spirit shone upon him, giving each individual thread in his coat of many colors a vibrancy, vitality, and vivacity that is unique to holy vestments. The colors in his coat were fast and did not fade because he was diligent to properly care for and maintain it. Because he did not defile it, but was true and faithful to the care instructions that were clearly printed on its label that came to be known as the Standard Works, it became a shield of protection to him. To avoid having its powers of enchantment neutralized, Joseph was careful to never inadvertently or carelessly contaminate his coat with other garments that might have been soiled with the stain of sin.

Our coats, that are similar in design and purpose to Joseph Smith's, have many colors, and we are able to differentiate each one from its companions by its own unique characteristics. Psychophysicists tell us that the human eye can distinguish up to 10 million different colors, which is remarkable, since there are only three primary colors in the visible light spectrum (red, green, and blue). Isaac Newton, who was the first to use a prism to separate white light (at wavelengths between 390 and 700 nm) into its individual colors, divided the visible light spectrum into seven named colors (red, orange, yellow, green, blue, indigo, and violet). So, the arrangement of colors in our own coats leaves plenty of latitude for us to be unique, to fit our circumstances. The key that unlocks the power of the colors is our ability to recognize the strengths of each of them, and then use them to good purpose.

In general, the color red calls us to action, and reminds us that the Savior trod the winepress alone. Orange is a warning to take care that we conform our lives to the Lord's design. Yellow encourages us to seek the light that is gathering in the east. Green suggests the power of envy and demands that we observe the 10th commandment and are content with the cards in the hand that we have been dealt. Blue reminds us to mourn with those that mourn, and to comfort those that stand in need of comfort. Indigo is a color whose depth and brightness represent the profundity of the Gospel and its ability to illuminate truth wherever it may be found. Violet is the color of amethyst, lavender, and beautyberries, and reminds us of the garlands festooning the avenues of the celestial city of God. (See Revelation 21:20).

Grey (black and white) is associated with conformity, indifference, neutrality, and uncertainty. It prompts us to choose whom we will serve and encourages us to take the Lord's side on all issues. Purple (red and blue) urges us to remember the royal robes of Christ our King. Black (blue, red, and yellow) underscores the necessity of opposition that paves the way to our progression. White (red, orange, yellow, green, blue, indigo, and violet) solemnly suggests the totality of the ordinances of the priesthood, our temple covenants, and the purity of the Spirit, all of which are necessary if we are to regain the glory of our former home.

When we finally come face to face with extra-terrestrial beings, the encounter will hopefully be accompanied by a profusion of color that dazzles their eyes. For all we know, color may rank high on their list of emotionally subliminal messages that accompany First Contact. In anticipation of the festivities, we should wear, as it were, a visual tutorial to which our alien visitors might easily refer, to allay fear, suspicion, and mistrust. Under the best of circumstances, our coats will be as an artist's palette that conveys acceptance, benevolence, and warmth. They will reveal our hearts' desires, our true intentions, and, yes, our true colors. There will be no hidden agenda, no deception, and no holding back during First Contact. We anticipate that it will be an exciting, imaginative, intriguing, lively, vibrant, vivid, and even colorful experience for all.

From ultraviolet to infrared, the coats we wear incorporate into their pattern and design every color of visible light, but they also resonate with radiation from a spectrum that can only be seen with eyes that have been touched by the finger of God. If we were able to break down that energy with a spiritual prism, we would comprehend the displayed colors in a panorama of eternity. "By the power of the Spirit our eyes (would be) opened and our understandings (would be) enlightened, so as to see and understand the things of God." (D&C 76:12). Perhaps seeing the millions of colors of the visible light spectrum would unfold to our view the Pillars of Creation itself. The key of the mysteries of God, as well as of the cosmos, would expand the borders of our understanding of ourselves and the galactic neighbors we are so anxious to meet.

Viewing reality and communicating with others through the discerning lens of color helps us to see how extra-terrestrial beings might find it easier to relate to us based, not only upon spiritual similarities, but also on the shared somatic senses of smell, sound, touch, and taste, and even on a shared recognition of visual geometric and chromatic patterns. In this vein, we think of the shapes in the motion picture "Arrival" wherein aliens initiated communication in a language written with circular symbols called logograms, or in the motion picture "Close Encounters of The Third Kind, where aliens first "spoke" from the sky over rural Mexico. In their words of the peasants who witnessed the manifestation, "the sun came out at night, and sang to us."

Later in the motion picture, terrestrial scientists communicated with a UFO through the medium of light. Those on the alien ship broadcast a neon display of colors that was correlated to sound, and then the human scientists responded in kind. Finally, both species exchanged "Curwen" hand signals that corresponded to a repeated five note tonal phrase that had initially drawn the "Mother Ship" to a hastily improvised landing site atop Devil's Tower, in Wyoming. After the extra-terrestrials had replied with the same gestures, sounds, and colors, they smiled, and returned to their ship, which then ascended into the night sky. To all who viewed this enactment on the silver screen, it was a satisfying precursor to what many hoped would someday prove to be real word First Contact.

How will we prepare our minds for our first exchange with aliens? On Friday, September 13, 2002, the world of Jason Padgett, for one, changed forever. During a mugging in Tacoma, Washington, he suffered a traumatic brain injury. As he recovered, he realized that his view of the world around him had been radically altered. He spontaneously began to see mathematical relationships everywhere he looked. Even things as ordinary as brushing his teeth were now dictated by mathematics. He repeatedly dipped his toothbrush in the tap water exactly 16 times, saying ""I like perfect squares. it's two to the power of four, or four squared." Clearly, neural connections in his brain had been rewired.

Water coming down the drain didn't flow smoothly anymore. Instead, it looked like geometrically perfect tangential lines. He stumbled across a webpage about fractals which struck a chord with him, and he started drawing them spontaneously and to perfection. Fractals are complex patterns that retain their similarity across different scales, in an endless feedback loop. They are represented in nature as trees mountains, clouds, and seashells.

Padgett's appreciation of fractals required him to master a difficult mathematical concept, which can be likened to conceptualizing and drawing snowflakes. When Padgett zoomed in with his mind, he saw that the fractals were made up of smaller "snowflakes" that were all connected. He visualized smaller and smaller perfectly symmetrical designs, in an endless downward digression.

Eventually, Padgett was diagnosed with acquired savant syndrome and a form of synaesthesia. Some synaesthesiacs see certain colors when they hear music, or their sense of smell is triggered when feeling a particular emotion. Associations can occur in any number of senses or cognitive pathways. The condition is caused by poorly understood neural connections. Some synaesthesiacs are born with the condition, while in others, such as in Padgett's case, their brains have been transformed by trauma, injury, or stroke.

Padgett's brain allowed him to visualize mathematical formulas and geometric shapes, both in his mind and in projections. He had inadvertently gained access to parts of the brain that are normally "off-limits", and his visual cortex began working overtime and in concert with the part of the cerebral cortex that does mathematics. He now sees beauty everywhere, and he is mesmerized by simple things that escape the notice of "normal" people, For Padgett, raindrops are transformed into complex overlapping rippling patterns that form shapes like stars.

Padgett's newfound perceptions have profoundly changed his view of reality. He is constantly amazed by the geometric harmony and beauty of the world, something he had beforehand taken for granted. "I'm having a mathematical awakening," he said, "and magic is all around me." Since his diagnosis, Padgett has published a book about his experiences entitled: "Struck by Genius."

To other synaesthesiacs, time and numbers are arranged with mathematical precision in physical space. Days, weeks, months, years, and even centuries have spatial forms; some appear closer in perspective than others. Colors, smells, or sounds are associated with written words. Numbers may have a specific size or shape. For example, people's ages may be expressed as curved lines. The Nobel Prize winning physicist Richard Feynman, who was a synaesthesiac, saw the letters in equations as colors, visualized dazzling quantum interactions, and correlated the intensity of their brightness to their mathematical significance.

The question is: Might it be possible to train ourselves to see the world as synaesthesiacs do, from a multisensory perspective? Might extra-terrestrials bring synaesthesia to the table, as well? Might that perspective enhance the probability of a successful alien First Contact? All of us might even now have these abilities, in varying degrees. For example, when we see red and blue flashing lights in our rear-view mirror, our respiration and pulse rates increase, and our pupils dilate. We have a measurable physiological response to a specific external stimulus that is based on color. While driving, and we hear a siren behind us, we automatically slow down and pull over to the right. We associate hearing with learned behavior. If we are in a building and we smell smoke, we nervously look for a fire exit. Sense and perception work in concert to protect us from harm. If we touch an electrical wire and feel a buzz, we quickly jerk our hand away. Our somatic senses enhance our survival instincts, and if we are not quick learners, we don't survive. If we take a drink of milk, and it tastes sour, we quickly and almost reflexively spit it out. We are discriminating in our taste, and characteristically and intuitively know how to avoid gastric distress. All these sensations might leave us with an unpleasant after-image that affects our future behavior when facing similar situations.

A related question is: For the same reasons, might extra-terrestrial beings have a different sense of how they relate to their environment? Are their brains wired differently? Do they exhibit the behavior of synaesthesiacs? Might synaesthesia help to ease our transition during First Contact from an anthropomorphic perspective to a more all-encompassing interstellar perspective that is more "real"?

To qualify as a synaesthesiac, the effect needs to be conscious, consistent, and automatic. No one is sure what causes synaesthesia, although it does seem to run in families. This is known as developmental synaesthesia, and it differs subtly from injury-related synaesthesia that is the result of brain "damage." Interestingly, synaesthesia is more common in those who are on the autism spectrum. So, "here's to the kids who are different, the kids who don't always get A's, the kids who have ears twice the size of their peers, and noses that go on for days. Here's to the kids who are different, the kids they call crazy or dumb, the kids who don't fit, with the guts and the grit, who dance to a different drum Here's to the kids who are different, the kids with the mischievous streak. For when they have grown, as history has shown, it's their difference that makes them unique." (Anonymous).

Terrestrial explorers who have ventured onto the unfamiliar terrain of time and the mind-expanding vistas of space might have been aided by synaesthesia. Following his voyage of discovery, Christopher Columbus wrote: "The Spirit gave me fire for the deed!" He might have written: Synaesthesia propelled me on my journey." Albert Einstein, recalling the mental gymnastics that were required as he wrestled with the equations related to his theories of relativity, wrote: "A storm broke loose in my mind." (Or, "synaesthesia expanded my awareness.")

There's evidence that some aspects of synaesthesia can be learned, but there's more to it than early childhood memories. Otherwise, many people in the Western world would likely associate 'M' with the color yellow, thanks to McDonald's ubiquitous golden arches. No-one knows why these color associations take root in some people and not in others.

Scientists are still debating how much of synaesthesia is genetic and how much is learned. While it appears that a typical person would struggle to train their brain to see a multi-sensory world as natural synaesthesiacs do, there's evidence that elements of the condition can be acquired. So, if we believe our lack of synaesthesia is holding us back creatively, maybe we should try writing our ideas out in colorful pens, or even with children's refrigerator magnets. Or, maybe we should choose different text colors, or try using different fonts, when typing word documents on our computers.

At the end of the day, synaesthesia might help ease our transition to membership in the United Federation of Planets, which in the Star Trek Universe boasts over 150 sentient species as members. Think of it: Each has lent their biological and technological distinctiveness to the Federation, without sacrificing their individuality. For all we know, synaesthesia is endemic among the members of the Federation, and it is this condition that has facilitated their integration into that interstellar union of brothers and sisters.

Chapter Eleven

The Universe is a Star Nursery

Has our Creator fashioned a universe, worlds without end, according to a Divine Design? Is our galaxy not only a star nursery, but also a neonatal stellar incubator, and a pediatric intensive care unit for protostars?

The Lord has declared: "My works have no end, neither beginning." (D&C 29:33). "And God spake unto Moses, saying: Behold, I am the Lord God Almighty, and Endless is my name." (Moses 1:3). Among His limitless creations, how many stars must be born, to help Him bring to pass His work and glory? We can only guess, but consider this: There are 600 billion galaxies in the known universe, and, within each one, three stars are born per year. (Best estimate, by Neil deGrasse Tyson, and other physicists). At the same time, they suggest that one star "dies" per year.

That is +3 and -1 = a net gain of 2 stars. But, to be on the conservative side, let's say that just 1 star per year, and not 2, is added each year to the celestial inventory within a typical galaxy, among the six hundred billion galaxies in the known universe. (These calculations were published in "Astrophysical Journal", based on observations of the Hubble Space Telescope, (Deep Field series), and with NASA's New Horizons spacecraft. Data suggest there may be between two hundred billion and a trillion galaxies. The calculations in this chapter use the average of NASA's estimate – six hundred billion galaxies). Therefore, the cumulative number of stars born every year in the known universe might plausibly be six hundred billion: 1 new star per year, per galaxy. That is roughly 1.6 billion new stars per day. (Actually, 1,643,835,616 new stars per day).

In the past 413 years, ever since Galileo Galilei heard about the "Danish perspective glass" in 1609 and looked up at the stars through his own primitive telescope that he built according to its specifications, over two hundred forty-seven trillion two hundred billion stars have been born. While you have been reading this chapter, (for the past four minutes), four million five hundred sixty-six thousand two hundred and ten stars have been born. In the second that it has taken to process that number with your brain, another nineteen thousand and twenty-five stars were born. When you go to bed tonight, under a waxing moon and a starlit sky, by the time you wake up in the morning after a refreshing eight hours of sleep, (which is eight hours of active labor in the birthing center of the universe), another five hundred forty-eight million stars will have been born. (547,945,205, to be exact – that's over half a billion stars!) We can scarcely keep up, but suffice to say: There is plenty of room in the universe for extra-terrestrial civilizations to flourish. We can take the Lord at His word, when He declared: "And worlds without number have I created; and I also created them for mine own purpose." That purpose is to be "a machine for the making of gods." (Henri Bergson, "Two Sources of Religion and Morality").

Chapter Twelve

If You Could Hie to Kolob

"The Lord's throne is in heaven," wrote the Psalmist. (Psalms 11:4). In the beginning, when God created the heaven and the earth, He made them temporally and spatially separate, with unique bounds and conditions. It was this stroke of genius on the part of our Father that allowed Him to manipulate the laws of physics to create a veil, as it were, so that we would forget all about our pre-mortal home, that we might take full advantage of our mortal condition, in harmony with His Merciful Plan.

Nevertheless, we do know something about our place in the cosmos, heaven, and the eternities, because according to the Book of Abraham's Facsimile #2, a place exists that is named Kolob, signifying the first creation, nearest to the celestial, or the residence of God. Of our relationship to that realm, William W. Phelps wrote: "No man has found pure space, nor seen the outside curtains, where nothing has a place." In the matrix of the dimensional reality in which he envisioned Kolob, there is no end to matter, space, spirit, or race, virtue, might, wisdom, or light, union, youth, priesthood, or truth, glory, love, or being. (See "If You Could Hie to Kolob"). These are defined by bounds and conditions that are only comprehensible to God.

Ultimately, said the Lord, "there shall be the reckoning of the time of one planet above another, until thou come nigh unto Kolob, which Kolob is after the reckoning of the Lord's time, which Kolob is set nigh unto the throne of God, to govern all those planets which belong to the same order as that upon which thou standest." (Abraham 3:9). Somehow, it is from Kolob that the order of the other creations of God is temporally and spatially governed, and it is from there that the boundary of heaven is established in such a manner that it is beyond the reach of detection by even the most sophisticated and delicately calibrated instruments utilized by terrestrial scientists. For example, the Hubble telescope can see billions of light years into our past, almost back to the moment of creation at the Big Bang, but it cannot gaze into heaven for five minutes. If it could do that, we "would know more than (we) would by reading all that has ever been written on the subject." (Joseph Smith, H.C. 6:50). Especially in the case of higher temporal and spatial dimensions, some things need to be believed to be seen.

Isaiah confirmed the temporal and spatial distinctiveness of heaven and earth. "The heaven is my throne," the Lord revealed, "and the earth is my footstool." (Isaiah 66:1). It is the Spirit, however, that has the power to carry us beyond the perceptible and palpable confines of this world to a place where boundaries become blurred, and the barricade of borders disappears. As John the Revelator exclaimed when he received his apocalypse: "Immediately I was in the spirit, and behold, a throne was set in heaven." (Revelation 4:2). Joseph F. Smith had a similar experience, when "the eyes of (his) understanding were opened, and the Spirit of the Lord rested upon (him)," and he, too, saw into the eternal world. (D&C 138:11). Normally, the veil functions as an event horizon that denies our senses any hint of what lies beyond. Only the Spirit has the power to penetrate the barrier that isolates us from the sum and substance of eternal reality. Only the Spirit will answer our questions: "O God, where art thou? And where is the pavilion that covereth thy hiding place? (D&C 121:1).

In the beginning, it was "the Gods (who) organized and formed the heavens and the earth" by defining the boundaries of the temporal universe, not to mention the conditions that govern the eternal world. (Abraham 4:1). They did this by the power of faith. They set the conditions "by which the worlds were framed, (and) all things in heaven, on the earth, or under the earth. (These) exist by reason of faith as it existed in (the mind of the Gods). Had it not been for this principle of faith, the worlds would never have been framed, neither would man have been formed of its dust. It is this principle by which Jehovah works, and through which he exercises power over all temporal as well as eternal things." (Joseph Smith, "Lectures on Faith", #1). Perhaps, this is why it is only by exercising perfect faith that we can have a true understanding of the universe and begin to experience the stability of God's reality. (James 2:22). Ultimately, "truth is knowledge of things as they are, and as they were, and as they are to come." (D&C 93:24).

Physics tells us that there are no privileged frames of reference. The galaxies are imbedded in time and imprinted upon space, within a fabric that is constantly expanding. If we ask where and when the creation took place, the answer is everywhere and forever. If the universe is warped through time and space into multiple dimensions, it might expand much like a balloon, creating in every instant more space. It seems reasonable that God would utilize our everyday laws of physics to accomplish these purposes beneath the umbrella of the eternal thrones, dominions, principalities, and powers that are the cornerstones of His own familiar higher-dimensional reality. This may explain why the Lord said to Moses: "As one earth shall pass away, and the heavens thereof, even so shall another come, and there is no end to my works." (Moses 1:38). "For by him were all things created that are in heaven, and that are in earth, visible and invisible." (Colossians 1:16).

For now, our poor lenses have difficulty discerning what is really there, for "no man hath seen God at any time in the flesh, except quickened by the Spirit of God." (J.S.T. John 1:18). If "the light of the body is the eye," then, when the eye is single to faith, our "whole body shall be full of light." (3 Nephi 13:22). On one occasion, after having received revelation, Joseph Smith confirmed the veracity of that promise, declaring: "My whole body was full of light, and I could see even out at the ends of my fingers and toes." (N.B. Lundwall, "The Vision," p. 11). This may be why the angel Moroni hovered in the air during his visits to Joseph Smith in his chamber, and why his hands and his feet were naked. (J.S.H. 1:31). He wanted to "see" with every part of his body. Every child of God potentially possesses this gift, and the Lord has promised that it only waits to be revealed. "If your eye be single to my glory," He said, "your whole bodies shall be filled with light, and there shall be no darkness in you; and that body which is filled with light comprehendeth all things." (D&C 88:67). There will come a day when "the sun shall no more go down; neither shall (the) moon withdraw itself, For the Lord shall be (our) everlasting light." (Isaiah 60:20).

When each of us comes face to face with eternity, as we must, a spiritual element will transform our mortal clay. Beforehand, while we tarry on the Earth, we might ask under what circumstances might that element quicken us, and how can the pure knowledge that flows out of it be vitalized? "A man's wisdom maketh his face to shine, and the boldness of his face shall be changed." (Ecclesiastes 8:1). When we are at one with God, when we have spiritually been born of Him and have internalized His divine nature, we will receive His image in our countenances. (Alma 5:14). That image and His likeness will bridge the barriers of time and space to leave their mark as an undeniable confirmation of our noble birthright. Our genetic code will be manifest as it influences all of nature; it will transform us as it blesses us with an endowment of unearthly powers.

When we move into eternity, time will lose all significance, and "See you later" will cease to be in our vocabulary. Time, that we too frequently viewed as a predator that stalked us all our lives, will then be fondly remembered as the companion that accompanied us on our journey through mortality, reminding us to cherish every moment. But more than that, we will discover that our mortal walk was only a tiny fraction of a much longer journey, and that our perspective was faulty only for as long as we believed it to be unique.

We may be shocked when we learn that mortality was not our natural dimension, after all. We will come to understand why it was that we were never entirely comfortable in time, and why we felt like "strangers and pilgrims on the earth." (Hebrews 11:13). This will, in turn, explain our innate thrust always toward the future, always beyond the horizon, and always upward, toward the stars. It will explain our intrigue with the heavens, and why our interstellar fascination had been so powerful.

For now, while we remain trapped in time, we can only indirectly appreciate the eternities. As we seek learning, even by study, and also by faith, "we can make our lives sublime, and departing, leave behind us footprints on the sands of time." (Henry Wadsworth Longfellow, "A Psalm of Life"). There is always the threat, however, that those footprints might be washed away by the incessant action of temporal and spatial waves beating on our shores. "There is a tide (after all) in the affairs of men which, taken at the flood, leads on to fortune. Omitted, all the voyage of their life is bound in shallows and in miseries. On such a full sea, (we are) now afloat, and we must take the current when it serves, or lose our ventures." (Shakespeare, "Julius Caesar," Act 4, Scene 2).

The Gospel is a practical guide for living in the here and now, but it also anchors us to the infinite and confirms that our destiny was prepared in the pre-earth existence, is molded in mortality, and will be established in eternity. When we pass through the veil, the heavens will smile upon us, and we will be clothed with the glory of God amidst an infinite hierarchy of temporal and spatial dimensions that finally renders coherent the Savior's promise that in His Father's house there are many mansions. (See John 14:2).

When we are spiritually sensitive, the veil becomes almost transparent. Then, as our powers expand, we experience the glittering facets of the life of the Spirit. We use the careful preparation and training we have received as a springboard, and we find ourselves in command of disciplined, controlled procedure. We are receptive to flashes of insight, as God's Plan liberates us to be creative, which encourages greater freedom. It is "the perfect law of liberty." (James 1:25). The Gospel amplifies the quiet spiritual stirrings that accompany and underlie our experience. It is the catalyst that has been added to a celestial cocktail that was designed to be savored as a heady elixir powerful enough to propel us into the presence of God.

"I wish I could remember the days before my birth, and if I knew the Father before I came to Earth," mused the poet. "In quiet moments when I'm all alone, I close my eyes and try to see my heavenly home. Although I can't remember and cannot clearly see, I listen to the Spirit and so I must believe. But still I wonder, and I hope to find the answer to the question that is on my mind. Where is heaven? Is it very far? I would like to know if it's beyond the brightest star." (Janice Kapp Perry).

Chapter Thirteen

"Me Transmitte Sursum, Caledoni!"

Transporters were used in the Star Trek Universe to convert persons or objects into an energy pattern (in a process of "dematerialization") that was then "beamed" to another location, where it would be "rematerialized." However, the familiar command: "Beam me up, Scotty" is a misquote, although in the original Star Trek, there were a number of near misses: "Mr. Scott, beam us up". "Scotty, beam us up." "Scotty, beam me up." "Beam them out of there, Scotty." and "Beam me up."

In the Star Trek Universe, the transporter was invented in the 22nd century by Dr. Emory Erickson, who was the first human to demonstrate the technology's practical applications. Within two centuries, trans-portation had become the safest way to travel, although the range was only 40,000 kilometers. Transport could be initiated at warp speed, but only if both vessels were traveling at the same velocity.

Transporters were fitted with Heisenberg Compensators that conveniently removed uncertainty from measurement of the subatomic particles that were dematerialized / rematerialized. They were also fitted with pattern buffers that permitted a degree of leeway when confronting the laws of physics. Additionally, transporter scanners could detect and deactivate weapons, and a biofilter removed attached contagions. Finally, a memory file of the pattern was created, in case anything went wrong during the transport.

The real-world problem with teleportation is that the matter in our bodies stubbornly obeys the immutable laws of physics, such as those relating to the strong and the weak nuclear forces. These are two of the most powerful bonds in nature. Governing laws prevent, for all practical purposes, turning human beings into so many zeroes and ones. If matter did not obey these laws, every atom in the universe would spontaneously explode, leaving it without structure in a chaotic mess. The nuclear forces act as superglue at the atomic level, holding both inert and biological tissues together. They resist anything even remotely resembling transporter technology.

Evidently, Jeremiah beheld the Earth as it was before those forces had been functionally applied to terrestrial objects, and he commented on the tumultuous conditions. "It was without form, and void; and the heavens … had no light. I beheld the mountains, and, lo, they trembled, and all the hills moved." (Jeremiah 4:23). Moses simply reported that before the moment of its creation, "darkness was upon the face of the deep." (Genesis 1:2).

Since the creation of the Earth, however, the forces of nature have held up remarkably well, with few exceptions. (See 3 Nephi 9). Today's technology has allowed physicists to tinker with nature to give up some of its secrets. (See Chapter Five). It is impossible to say whether advanced extra-terrestrial species have been able to surmount the formidable physical obstacles to teleportation. But if they have been successful, one day they might just be able to teleport themselves right into First Contact. If they chose instead to transport us to the party, it would be the ride of a lifetime.

So far, terrestrial scientists have successfully transported only small quantities of photons from one location to another, in a process known as quantum transport. The intended applications are in the field of electronics, such as advanced communications and quantum computers. Before we move any further exploring the feasibility of transporter technology, we should first address related philosophical and ethical questions. How would the system handle a transported person's consciousness, personality, memory, and identity? In what way might teleportation influence their moral behavior? Would it wreak havoc with their belief system and with their faith? Is teleportation related to being born again? Could we ever hope to re-materialize the 37 trillion cells in a human body in exactly the right arrangement, without subtly negatively disrupting molecules, tissues, organs, or organ systems, let alone our sum and substance?

It is possible that humans will yet discover unimagined breakthroughs in physics that would lead to the development of transporter technology. For the moment, however, all we can say is that it is not beyond our wildest imagination. Me transmitte sursum, Caledoni!

Chapter Fourteen

Does God Obey the Speed Limit?

As we attempt to wrap our minds around the concept of God, we need to realize that "nowhere is out of His presence, and now is forever. As time is no more, likewise space will shrink irrevocably. For all we know, the speed of light may prove to be too slow to do some of what must be done." (Neal A. Maxwell, "Ensign," 7/1982). For the time being, however, the laws of physics bind us to the cosmic speed limit, at least within the bounds of our solar system, the Milky Way galaxy, and the universe itself. In normal time and space, the speed of light is always the same, regardless of the velocity at which we may be traveling in relation to the referenced beam of light. Whenever anyone traveling at a steady velocity measures the speed of light in empty space, she always comes up with the same answer: 186,282.3969 miles per second, regardless of how fast or in what direction she may be moving.

This physical constant prompts the questions: "Might there be extra-terrestrials living in Type 5 Omega Minus civilizations, who can travel with impunity throughout the universe? (See Chapter Five: Charting Our Course Through the Cosmos). For that matter, does God obey the cosmic speed limit?" After all, yesterday, today, and forever are ever before Him in one eternal round, and He is "the Great I AM, Alpha and Omega, the beginning and the end, and the same which looked upon the wide expanse of eternity, and all the seraphic hosts of heaven before the world was made." (D&C 38:1-2). This also begs the questions: As the arrow of time flies through eternity, ostensibly linked to light speed, does it travel in two (or even multiple) directions, forward and backward at once and forever? If it does so, can the bonds of its spatial limitations and temporal constraints be broken, that seem to be etched in stone (at least here in the temporal world) at exactly 186,282.3969 miles per second?

Eternity is unique, although it shares surprising similarities to quantum mechanics' view of the universe. Both are open systems independent of spacetime that cannot be defined by temporal or spatial borders. Heavenly Father belongs to eternity, and so likely carries out His activities within the parameters of celestial laws that build upon, supersede, or simply do not apply to, the laws that govern our physical universe. As Paul explained to the Corinthian Saints: "The natural man receiveth not the things of the Spirit of God: for they are foolishness unto him: neither can he know them, because they are spiritually discerned." (1 Corinthians 2:14). A barrier exists for those of us with scientific, rational minds, that makes it difficult to consider the possibility that there might be a God in heaven who might not obey the cosmic speed limit that remains constant and unalterable in empty space.

We are more familiar with the variability of sound waves in our atmosphere, where their speed depends on air density, temperature, and humidity. But no such variations occur as light travels though the vacuum of space. At least superficially, we might conclude that this puts limitations on God. "But," we ask ourselves, "How could this be?" It is like asking the paradoxical question: "If

God can do anything, can He make a mountain that is too heavy for Him to lift?" Or, one could ask: "If God can do anything, can he establish a cosmic speed limit that He cannot exceed?" Or, it would be like saying to a member of the opposite sex: "If I ask you for a date, would your answer be the same as your answer to this question?".

It is not surprising that obedience to celestial laws allows us to more easily "see and understand the things of God, even those things which were from the beginning, before the world was." (D&C 76:12-13). Reconciliation with God allows us to escape the limitations of our mortal clay, as our minds transcend the temporally unalterable laws of physics, grapple with the spatially expansive laws of heaven, and finally expand in union with eternal principles that govern the Celestial kingdom. For now, though, we are somewhat like "a man beholding his natural face in a glass," satisfied with a fleeting glimpse of forever in a mirror, but without ever comprehending the true nature of eternity. (James 1:23).

The Lord is eager to share His knowledge of the laws that govern His creations with those who are prepared to receive it, for "eye hath not seen, nor ear heard, neither have entered into the heart of man, the things which God hath prepared for them that love him." (1 Corinthians 2:9). Because experiences with the Spirit defy rational explanation, it remains that no profane "tongue can speak, neither can there be written by any man, neither can the hearts of men conceive (the) great and marvelous things" that are easy for God to understand. (3 Nephi 17:7). He has given us our five natural senses that we might embrace our physical surroundings, but our familiarity with the things of the Spirit comes only from extrasensory perceptions that have the capacity to carry us beyond the conventional equations that define the mathematical boundaries of space and time.

On the Earth, when we measure a beam of sunlight streaming toward us, its speed is exactly 186,282.3969 miles per second. If we were to fly in a spaceship toward the Sun at half the speed of light and measure the speed of the same beam as it travels earthward, it would be clocked the same 186,282.3969 miles per second. If we were to turn the ship around and fly back home at half the speed of light, the same beam of sunlight overtaking us does so at exactly 186,282.3969 miles per second. No matter our speed or direction, that beam of light always stubbornly travels at the same speed, relative to our own velocity.

In the temporal universe, the speed of light is always the same relative to every observer, no matter how fast or slow they may be moving in relation to the light that is being measured. Nature accomplishes this trick because the judgments of space and time are "private" for each individual. Therefore, for those of us on the Earth, moving around its axis at 1,000 miles per hour, with it in orbit around the sun at 67,000 miles per hour, or moving with the solar system around the Milky Way at 137 miles per second, or moving with the Milky Way through intergalactic space at 1.3 million miles per hour, and another person were on a spacecraft traveling somewhere through the cosmos at a velocity independent of measurement by the Earth, the solar system, or the galaxy, you would both still get the same answer when calculating the speed of the same beam of light. It would be exactly 186,282.3969 miles per second. The simple explanation is that time slows down as we move faster. We will always get the same result, when measuring that beam of light, no matter how fast or slow we ourselves may be moving, because time is relative to our velocity. This is termed "time dilation."

Gradually, it dawns on us that Gospel principles relating to the eternities just might supersede physical laws relating to the temporal universe. Those who are "born of him" may be oriented more to the expansive laws of the eternal world than to the restrictive laws of the physical world. "Whatsoever is born of God overcometh the world" and is therefore free of the confinements of the equations of mathematics and the limitations of the laws of physics in an inexplicable, indescribable, and yet undeniable way. (1 John 5:3).

It is the Theory of Relativity that revealed that there are no privileged frames of reference. The galaxies are attached to space and imbedded in time, while its fabric is constantly expanding. Interestingly, theories have been postulated that fit the observable universe only back to a point just after its creation. Observational and experimental physical phenomena have been mathematically mapped out back to a few milliseconds this side of the theoretical Big Bang. At that point, just after the moment of singularity, the laws of physics break down, and we are left in the dark (no pun intended). This suggests that under different circumstances that fall outside the parameters of the observable universe, other metaphysical laws might apply. God works in mysterious ways, His wonders to behold. (William Cowper).

Ultimately, if we substitute the term "Creation" for "Big Bang" and ask where and when it took place, the answer is everywhere and forever. No one can say whether God utilized the laws of physics (as we understand them) during the Creation, but what we do know is that "by him, were all things created, that are in heaven, and that are in earth, visible and invisible, whether they be thrones, or dominions, or principalities, or powers." (Colossians 1:16). This leaves the door ajar for theologians to confidently debate science from a position of strength, and it goes a step beyond Intelligent Design, because it boldly testifies that God Himself was in control of the Creation. But if that is true, it also begs the question: "Does He need to obey the speed limit?"

When we transition into eternity, the constraints of time and space associated with the cosmic speed limit will lose all significance, and "See you later," will cease to be in our vocabulary. When time is no longer an element of the equation, by an act of God's Celestial Congress the speed limit will be effectively repealed because the reality in which it had been anchored will have been superseded by the mathematics of eternity. In a Land Without Time, the laws of physics will have lost all relevance.

Time, that we too frequently had viewed as a predator that stalked us all our lives, will then only be vaguely remembered as a companion that accompanied us on our journey through mortality, reminding us to cherish every moment.

Somewhere beyond time, when we confront eternity, we will find that the perspective we had believed to be unique was faulty. We will be surprised to learn that mortality was not our natural dimension. We will understand why we were never entirely comfortable in our mortal circumstances, and why we felt like "strangers and pilgrims on the earth." (Hebrews 11:13). In turn, this will explain our innate thrust always toward the future, always beyond the next horizon, and always toward the stars. In a coming day, we will understand the nuance between space and time, and realize that they are elements of a continuum that can actually be controlled. (See D&C 121:36).

For now, we remain "at work with our hands to the plough and our faces to the future. The castle of enchantment is before us, and daily we catch glimpses of its battlements and towers. The best of life is always further on. Its real lure is hidden somewhere behind the hills of time." (Sir William Mulock).

When we discover that castle of enchantment behind the hills of time, we will find that growing "older" at the rate of one day every twenty-four hours had been strictly and uniquely a quality of mortality and a brilliant mechanism designed by Heavenly Father to afford us an opportunity to gauge the approach of our reunion with Him in the eternal world. We will discover that we lived out our lives in only one dimly lighted corner of reality. Thus, it was difficult for us to appreciate our potential and grasp the power of our position, that we might one day "flourish in immortal youth, unhurt amidst the war of elements, the wreck of matter, and the crash of worlds." (Joseph Addison, "Cato," Act 5, Scene 1).

From the perspective of eternity, we will realize that we had been frozen in time, in the sense that we had grown complacent in our indifference to the subtle messages reflected in its passage. The numerical equations that had governed the temporal world will be to us as a child's counting game. "Tinker, tailor, soldier, sailor, rich man, poor man, beggar man, thief." And all the while, our minds had been shackled by limiting beliefs, and the opacity of the mathematics of the cosmic speed limit.

Our transition into eternity will thrust upon us the realization that time and space are unnatural dimensions in which we, as eternal beings, cannot be completely comfortable. We will suddenly realize that they are transitory, by definition, and it might have been an illusion that made it seem that it was we who moved through them, when it was really the other way around. We will eventually run out of time, and we will no longer be confined by space. In our youth, time never seemed to pass quickly enough, and we wanted "our own space." Perhaps we were so recently removed from the eternal world, that we were impatient to return to the familiarity of its more natural dimensions. In any event, we might look back and realize that as we approached the terminator line between mortality and immortality, our perception of space and of the passage of time changed again; the former shrank while the latter seemed to speed up.

The following anecdote might help us to better understand the relativity of time. Albert Einstein suggested "When you sit with a pretty girl for two hours, you think it's only a minute. But when you sit on a hot stove for a minute, you think it's two hours." ("New York Times", 3/1929).

For now, we occupy three-dimensional space and move in one direction through time at the precise rate of one solar day in every 24 hours. We can be certain that the boundaries of the land, seas, and skies, and the sun, moon, and stars will continue to be numerically defined by physical scientists with greater and greater precision, even as the existence and significance of these phenomena are esoterically debated by theologians and philosophers alike. But we also have the assurance that the relationship between mathematical boundaries and the metaphysical thrones, dominions, principalities, and powers that relate to the heavens will be revealed when the Spirit opens the eyes of our understanding to undreamed of vistas of otherwise inaccessible experience. For example, how can we now hope to comprehend the scope of Moroni's promise that "by the power of the Holy Ghost (we) may know the truth of all things?" (Moroni 10:5). One day, the truth will set us free from the limitations of ignorance so that we may be as one with the majestic clockwork, "like a bird that, pausing in her flight a while on boughs to light, feels them give way beneath her and yet sings, knowing that she hath wings." (Victor Hugo). The depth and breadth of our comprehension will finally put to rest the debates that have preoccupied the sages since the dawn of The Age of Reason. We will soar to new heights as science and religion reach a meeting of mind and spirit to clarify and simplify our understanding of our place in the universe and in the eternities.

The repeal of the comic speed limit will create new opportunities when the chasm between the physical and eternal worlds is bridged and "old things are passed away (and) all things are become new." (2 Corinthians 5:17). Paul repetitively described this as a process of "reconciliation." Ultimately, he wrote, "all things are of God, who hath reconciled us to himself by Jesus Christ, and hath given to us the ministry of reconciliation. To wit, that God was in Christ, reconciling the world unto himself." (2 Corinthians 5:18-19).

Perhaps, the Atonement, in its infinite complexity and awe-inspiring scope, provides a present-day exemption from obedience to the cosmic speed limit. When we are completely at one with Christ, our reconciliation to His perception of reality will allow us to be more easily governed by the laws that define how He relates to eternity. When we are at one with Him, we sense a freedom from incarceration to the will of the flesh. We may very well qualify by worthiness to travel together on a goodwill tour of His creations, with time slowing down as we move faster and faster, until we reach the speed of light, when it will be no more, and we are at one with the Infinite.

When eternity bursts upon us, we will be dazzled by an expanded appreciation of our relationship with Christ. It may only then dawn upon us that when He characterized Himself as the Light of the World, He may have been speaking literally as well as figuratively. In the physical world, time shifts compensate for our velocities relative to beams of light. In the eternal world, time shifts so completely in relation to the Light of Christ, that it is no more, and the cosmic speed limit becomes a negligible influence. Thanks to Einstein, we realize it's all relative! Thanks to Heavenly Father's utilization of eternal law, there is reconciliation between time and being that erases the cosmic speed limit and eases our transition to the eternities.

Heavenly Father shed light on this principle of reconciliation. "There are many kingdoms," He explained, "for there is no space in the which there is no kingdom, and there is no kingdom in which there is no space, either a greater or a lesser kingdom. And unto every kingdom is given a law; and unto every law there are certain bounds also and conditions." (D&C 88:37-38). In other words, there are many realities, or kingdoms, and every one of them occupies space.

But implied in this explanation is the suggestion that each space may be governed by certain bounds and conditions specifically tailored to its individual circumstances. God rules in both "the heavens above, and in the earth beneath" according to the bounds and conditions that are unique to each of those dominions. (Abraham 3:21).

When Jehovah stood in the presence of God the Father at the time of the creation of the Earth, He said to those assembled: "We will go down, for there is space there, and we will take of these materials, and we will make an earth." (Abraham 3:24). The space reserved for the Earth already existed. All that was necessary was for Jehovah to "go" there and establish, or set in motion, the laws, bounds, and conditions by which it could exist as a temporal entity in the three-dimensional space and one-dimensional timeline that is our present reality. As Luke wrote, God "made the world and all things therein ... and hath determined the times ... and the bounds of their habitation." (Acts 17:24 & 26). He established both the spatial and temporal conditions that would define earth life within our universe. He established the cosmic speed limit but inserted a codicil. Eternal progression would activate temporal and spatial exceptions that would forever allow Him, and that would one day allow us, to exceed the limit with impunity.

We are generally locked on telestial targets and content ourselves to be governed by a rev-limiter on the power plant that fuels both light bulbs and stars. This controls both the output of photons and their velocity of 186,282.3969 miles per second. We are drawn to the light like moths to fire, but we flutter around without purpose and direction. This means that higher-level thinking is too often beyond the reach of our comprehension. This is because, as the Lord reminded us: "My thoughts are not your thoughts, neither are your ways my ways ... For as the heavens are higher than the earth, so are my ways higher than your ways, and my thoughts than your thoughts." (Isaiah 55:8-9). We cannot penetrate the mind of God unless He allows us to do so.

Left to our own devices, we will never understand the physical universe so completely that we will become its master, the possibility notwithstanding of our achievement of Type 4 or 5 civilization status. (See Chapter Five: Final Thoughts). We cannot hope to supplant God's intellect with our own. We can never, in any significant way, understand the eternities while we remain in the stew of seconds, the mire of minutes, and the agony of hours in the physical world. Well did the Psalmist counsel: "Be still, and know that I am God." (Psalms 46:10). For now, we just need to plod along at or below the posted speed limit that provides us with order and gives us peace of mind. We need to be content to labor in the traces.

But "great and marvelous are the works of the Lord," Jacob exclaimed. "How unsearchable are the depths of the mysteries of him; and it is impossible that man should find out all his ways. And no man knoweth of his ways save it be revealed unto him." (Jacob 4:8). "O the vainness, and the frailties, and the foolishness of men!" wrote Lehi. For "when they are learned, they think they are wise" and they suppose "they know of themselves, wherefore, their wisdom is foolishness and it profiteth them not." (2 Nephi 9:28-29). As Paul cautioned the Colossian Saints: "Beware lest any man spoil you through philosophy and vain deceit, after the tradition of men, after the rudiments of the world." (Colossians 2:8).

We don't understand how God accomplishes the trick of moving throughout His universe impervious to the speed of light, but we know that at ludicrous speed corruptible matter turns into pure energy. One might say that it turns into the power of God, that is even greater than the Omega Particle of Star Trek lore. (See Chapter Five: Charting our Course Through The Cosmos). One could even say that His energy = MC2, or mass times the speed of light squared. This is a very large number, indeed! If you were traveling very near the speed of light, it would take a nearly infinite amount of energy to increase your speed by just a few feet per second, to actually reach the cosmic speed limit, or the speed at which the mind of God kicks into high gear. This is why "Spaceballs," the motion picture spoof of "Star Wars," utilized "Ridiculous Speed," and "Ludicrous Speed." However, "with God, all things are possible." (Matthew 19:26).

The theoretical physics behind "warp drive" engines may help to explain how mere mortals could meet the energy requirements of light speed. In the Star Trek Universe, these drives generate incomprehensible amounts of power, on the order of billions of gigawatts. (See: TNG: "True Q"). But more fundamentally, in Star Trek, when a ship goes to warp, it moves through space at speeds faster than the speed of light, but only because time has been warped. The ship has taken advantage of the curvature of spacetime and has (indirectly) gotten from point A to point B at a velocity faster than the speed of light. The distance between those two points is no longer a straight line. Essentially, the ship takes advantage of a warp bubble, to bunch up space in front, and stretch it out behind the vessel.

In the real world, NASA (National Aeronautics and Space Administration) and DARPA (Defense Advanced Research Projects Agency) have set a goal to have the United States reach "warp capability" by the end of the Twenty-first Century. These agencies envision faster-than-light travel, except it's not moving faster than the speed of light – it's more like bending the fabric of spacetime while you stay still.

To better understand the concept, visualize a string held taught between your two hands. On the string, an ant is traveling at the speed of light. If your hands are brought together so that the string hangs down and its two halves touch, the ant will be able to easily move from a point on the first half of the string to the same point on the other half. If the string is then returned to its taut position, the ant will have miraculously moved a significant distance in spacetime, in an instant. In this example, faster-than-light "speed" will have been accomplished, but not by straight line travel from one place to another.

In this sense, Star Trek "Warp Factors" do not really measure speed as much as they measure to what extent space can be warped. If Warp Factor 1 is distance traveled at the speed of light, then Warp Factor 2 would be a measure of the distance traveled as if one were moving at twice the speed of light, and so on.

For an in-depth journey through the physics of warp bubbles, go to" The European Physical

Journal C", (2001) 81:677, "Worldline Numerics Applied to Custom Casimir Geometry Generates Unanticipated Intersection with Alcubierre Warp Metric"). But, be forewarned: The contents of that paper are not for the faint-of-heart.

What would the design of a warp-capable ship be like? A round, saucer-shaped vessel, like those popularized by science fiction, might work in concert with a warp bubble, if one snag in the theory of time travel can be solved. The problem is that if a ship is traveling faster than light, it goes through time more slowly than those of us (outside the bubble) who are obeying the cosmic speed limit. For those on the ship, time seems normal. But compared to those on a planet, or a space-station, for example, ship time would lag far behind. If it were somehow possible to maintain normal spacetime within the Warp Bubble, perhaps the timeline, relative to those outside the bubble, could be preserved, as well. But, if not, chaotic temporal distortions would run amok.

For now, how might space-faring extraterrestrials deal with the problem of a temporal paradox? Imagine that a starship leaves its base at point A, and travels through space to points B and C within a warp bubble at faster-than-light speed. It then returns to base, at point A. When it arrives, it will be in spatial alignment with the base, but its temporal alignment will have been forever altered. It will be out of synch with those who had been left back at the base, because time will have passed more slowly for those on the starship, in relation to those who had remained behind at the base, in normal spacetime. Star Trek writers apparently solved the temporal distortion paradox with the stroke of a pen. In the Voyager episode entitled "Year of Hell," on of the characters declared: "The past, the present, and the future – they exist as one. They breathe together."

In that episode, an alien named Annorax attempts to create intentional temporal incursions in order to erase existing events from the historical record, through the utilization of a "temporal core" weapon. His intention was to restore the original timeline, but he did not realize that this could only be done by the destruction of the temporal core itself.

That may work in the Star Trek Universe, but we don't know how God will solve the temporal distortion paradox in the real world. We only know that He will. For now, He has only reassured us that "He hath given a law unto all things, by him they move in their times and their seasons. And their courses are fixed, even the courses of the heavens and the earth, which comprehend the earth and all the planets. And they give light to each other in their times and in their seasons, in their months, in their years – all these are one ... with God, but not with man." (D&C 88:42-44). For now, we are admonished: "Listen to the voice of the Lord your God, even Alpha and Omega, the beginning and the end, whose course is one eternal round, the same today as yesterday, and forever." (D&C 35:1-2). He 'doth not walk in crooked paths, neither doth he turn to the right hand nor to the left ... therefore his paths are straight, and his course is one eternal round." (D&C 3:2). It will not be too long before He "gather(s) together in one all things, both which are in heaven, and which are on earth." (D&C 27:13).

In the Star Trek Universe, in the seventh season of "The Next Generation", in an episode appropriately entitled "Forces of Nature", it was discovered that Federation starships traveling at warp speed had been creating subspace rifts that were catastrophically damaging the fabric of normal spacetime. As a result, the Federation immediately implemented an emergency speed limit of Warp 5.

Interestingly, in the real world, subspace has been theorized to exist in spacetime. String theories posit anywhere from 10 dimensions to 26 dimensions of spacetime to maintain mathematical consistency. If spacetime exists in these dimensions, it would include the three spatial dimensions and the one temporal dimension with which we are familiar, plus a number of others that are curled up on a subatomic level.

In the Star Trek Universe, subspace rifts were an anomaly, first encountered by the USS Enterprise, NCC-1701-D, in 2370, when the fabric of spacetime was weakened by high levels of warp energy. A cascade ensued, with tetryon radiation producing distortion waves that degraded the ship's shields. Had the Enterprise not immediately dropped out of warp, exposure would have caused a hull breach. Warp drive, at least in the short term, was rendered useless, due to its harmful effects within the rift. It was discovered that the continual use of warp drive could not only inflict heavy damage on the starship itself, but also cause the rift to further expand.

The Enterprise experience compelled starship engineers to create warp engine designs that utilized variable warp fields, specifically with the introduction of geometric nacelle technology. This essentially nullified the problem of damage to subspace by warp drive engines and allowed for the construction of even faster ships, such as Voyager.

In the real world, do we really need to understand the physics of warp drive, or might we expend our powers of inquiry more fruitfully to acknowledge the omniscience, omnipotence and even omnipresence of God? Soon enough, we will live in eternity, when time, and these questions, will be rendered insignificant. Then, we will better comprehend the expressions "Alpha and Omega," the "Beginning and the End," and "I Am that I am." When we move beyond mortality, (at faster-than-light speed?) time will cease to exist, and we will no longer need to wear wristwatches. "See you later!" and "Have a good day!" will no longer be in our vocabulary. For that matter, listlessness, boredom, monotony, and ennui will be alien to our experience. We will be up, and about, and moving. With Joseph Smith, we will exclaim: "O Lord God Almighty, maker of heaven, earth, and seas, and of all things that in them are ... let thy pavilion be taken up; let thy hiding place no longer be covered." (D&C 121:4).

We may one day find it ironic that the precious commodity of time itself had been the very source of our weariness and even of our impatience with life. We may discover that time was the very element that allowed our interest to wander. We might even find that we had fallen into the habit of killing time, rather than of making time, saving time, or finding time. We may one day expand upon Lehi's grand exposition on opposition to encompass both the positive and negative qualities of time. (See 2 Nephi 2:11). When we have overcome all things, and we are finally released from the constraints that had hindered our progress, enthusiasm and fervor may become our more natural passions. It may dawn on us that eternal progression is largely concerned with shedding the self-limiting and finite constraints of time from the unending equations that define improvement, permit progress, and stimulate growth and development.

One day, we may discover that mortality, with the arrow of time moving relentlessly, inexorably, and unremittingly forward in one direction was not our natural dimension. We may understand that it is from Kolob that the order of the other creations of God are spatially and temporally governed, and that it is from there that the boundaries of heaven are established. These are beyond the reach of detection by even the most sophisticated instruments utilized by physical scientists. The Hubble Space Telescope can see nearly 14 billion light years into our past, almost back to the moment of Creation at the Big Bang, but it cannot gaze into heaven five minutes. If we could do that, we "would know more than (we) would by reading all that has ever been written on the subject." (Joseph Smith, H.C. 6:50).

One day, we may realize that we never were meant to be entirely comfortable in our mortal circumstances, for the Earth is not our permanent home. We were born to ascend into heaven on chariots of fire. (See 2 Kings 2:11). We may come to realize that time itself was only a steady and methodical taskmaster that quietly motivated us to move forward with the covert encouragement of God. We may finally realize that time allowed us to live by faith. When we have run out of time and have overcome all things, and we are completely reconciled unto Christ and we are at-one with Him, we will no longer need the emotional crutch of time.

The temporal scope of God's kingdoms is beyond our comprehension, but prophetic insight provides clarity. Joseph Smith taught: "The great Jehovah contemplated the whole of the events connected with the Earth pertaining to The Plan of Salvation, before it rolled into existence, or ever 'the morning stars sang together' for joy; the past, the present, and the future were and are, with him, one eternal 'now.'" (Teachings," p. 220). The Savior has always existed in the present tense. His "course is one eternal round, the same today as yesterday, and forever." This allows him to see from multiple perspectives in time, "the beginning and the end." (D&C 35:1). As Alma explained to Corianton: "All is as one day with God, and time only is measured unto men." (Alma 40:8). Einstein was right; time is relative.

For the moment, we are trapped within the matrix of time, and seconds, minutes, and hours are the glue that binds us to the fabric of mortality. Thus, we can only indirectly appreciate the eternities. As we seek learning, even by study, and also by faith, "we can make our lives sublime, and departing, leave behind us footprints on the sands of time." (Longfellow). These footprints might be washed away by the incessant wave action of life's spatial and temporal dimensions, or perhaps by its distortions, beating on our shores. "There is a tide (after all) in the affairs of men which, taken at the flood, leads on to fortune." (Shakespeare, "Julius Caesar," Act 4, Scene 2). Our destiny was prepared in the pre-earth existence, has been molded in mortality, and will be established in eternity, when the heavens will smile upon us in a celestial moment that is frozen in time. The trajectory of our ascent will be breathtaking in its scope and mesmerizing in its brilliance. Our flight plan, that today obeys the cosmic speed limit, will then be unrestrained.

"The Lord's throne is in heaven," wrote the Psalmist. (Psalms 11:4). In the beginning, when God created the heaven and the earth, he made them spatially and temporally separate from each other. Their bounds and conditions were distinct. It was this stroke of genius on the part of our Father that has allowed us to forget all about our pre-mortal home so that His Merciful Plan could be fulfilled as we walk by faith, rather than by sight, here on Earth. Fortunately, the veil that was thus created solidly grounds us on the familiar bedrock of past, present, and future. Normally, it functions as an artificial horizon with which we are completely comfortable, and our senses give us no hint of what lies beyond. For all practical purposes, our present frame of reference allows us to live within a timeline that is woven into the tapestry of three-dimensional space. The arrow of time moves in only one direction, and it is deflected back into normal space whenever it bumps up against the veil. We are isolated by what appears to be, for all practical purposes, the sum of reality. It is only the unsettling influence of the Spirit that infuses us with the ability to see what lies beyond that event horizon.

The veil keeps things neat and tidy. It reassures us that the sun will come up tomorrow, and that twenty-four hours later it will do so again. The veil permits us to live by faith, even in our daily lives. Without it, life would be too confusing! When the veil that wraps us in time is no more, and after we have experienced a significant attitude adjustment, we will be free to put the pedal to the metal on the autobahn of life, racing through eternity with the top down and for the first time, perhaps, really feeling the celestial wind in our faces.

Then, we will enter the temporal and spatial reality of God's Rest, free of the self-limiting conditions that had previously blinded us to a larger view of life. When we are at-one with Him, we will realize that beforehand, we had been monotonously plodding along on our own, well below the posted speed limit. Surprisingly, His Rest will be as life in the fast lane, with no posted speed limit. His invitation to follow Him is even now prefaced by the action verb "come." Many times, the invitation is to "come, quickly." If we hitch our wagon to His star, perhaps we can even come to Him at or beyond the cosmic speed limit.

Perhaps, we can even "fall" into His Rest, for there is also a relationship between time and gravity. If you are standing on the edge of a black hole at the event horizon, you are on the edge of forever. If you fall in, or if a star system falls in, everything "stops" so to speak, relative to those still positioned on the "outside". Nothing can return, once it has fallen into a black hole. Not light, and not time. There is no going back.

Maybe this is what Joseph Smith meant, when he encouraged his brethren: "Shall we not go on in so great a cause? Go forward and not backward. Courage, brethren; and on, on to the victory!" (D&C 128:22). That victory may come when Christ reigns personally upon the earth and it is renewed to receive its paradisiacal glory. (See the 10th Article of Faith).

Some theoretical physicists believe that Black Holes might be wormholes to other places within our own universe, or even conduits to parallel universes. That may be true, but there is the problem of enormous gravitational forces that, it seems likely, would tear apart anyone unfortunate enough to enter one. That might not be an issue, however, for the inhabitants of Type 3, 4, or 5

civilizations who have learned how to harness the energy of the cosmos. Nor might it be an issue for resurrected, perfected beings who are much more resilient than carbon-based life forms. As the Lord told Moses: "No man can behold all my works, except he behold all my glory; and no man can behold all my glory, and afterwards remain in the flesh on the earth." (Moses 1:5).
There are probably many questions surrounding the creation of space and time that we will never be able to answer, as long as we remain imprisoned within our bodies of flesh and bone. The quiet reassurance, "In time, you will understand." may not be entirely satisfactory, for mortality itself impels us to see through a glass darkly. (See D&C 50:12). At some point when time no longer exists, we will be able to "go forth from our dwelling places and discarding the poor lenses of the body, peer through the telescope of truth into the infinite reaches of immortality." (Helen Keller).

But do these intellectual mind-games really have anything to do with our place in the cosmos? Do they relate to the questions: "Is anyone out there?" or "Are we alone in the universe?" Do they address the questions: "Where did we come from?" "Why are we here?" and "Where are we going?" That is anyone's guess. However, have you ever wondered, "Where is the center of God's universe?" "If it has an edge, what lies beyond?" "Where is heaven?" "Where are the kingdoms of glory of which Jesus spoke?" "Where is the spirit world?" "Where are the many mansions mentioned by the Savior?" "Where is outer darkness? "What is translation?" "How can God instantly hear and respond to our prayers?" "How does He perform all of the housekeeping responsibilities that are related to His kingdoms?" It is our God-given intellect that allows us to explore strange new worlds, to seek out new life, and new civilizations, and to boldly go where no one has gone before.

Of our relationship to His realm, William W. Phelps wrote: "No man has found pure space, nor seen the outside curtains, where nothing has a place." Perhaps in His kingdom, "there is no end to matter, space, spirit, or race, virtue, might, wisdom, or light, union, youth, priesthood, or truth, glory, love, or being," because these things are defined by different bounds and conditions in the infinite reaches of immortality and eternal life, where God and His exalted sons and daughters do not obey the speed limit. ("If You Could Hie to Kolob", W.W. Phelps).

"I wish I could remember the days before my birth, and if I knew the Father before I came to Earth. In quiet moments when I'm all alone, I close my eyes and try to see my Heavenly home. Although I can't remember and cannot clearly see, I listen to the spirit and so, I must believe. But still I wonder, and I hope to find the answer to the question that is on my mind. Where is Heaven? Is it very far? I would like to know if it's beyond the brightest star." (Janice Kapp Perry).

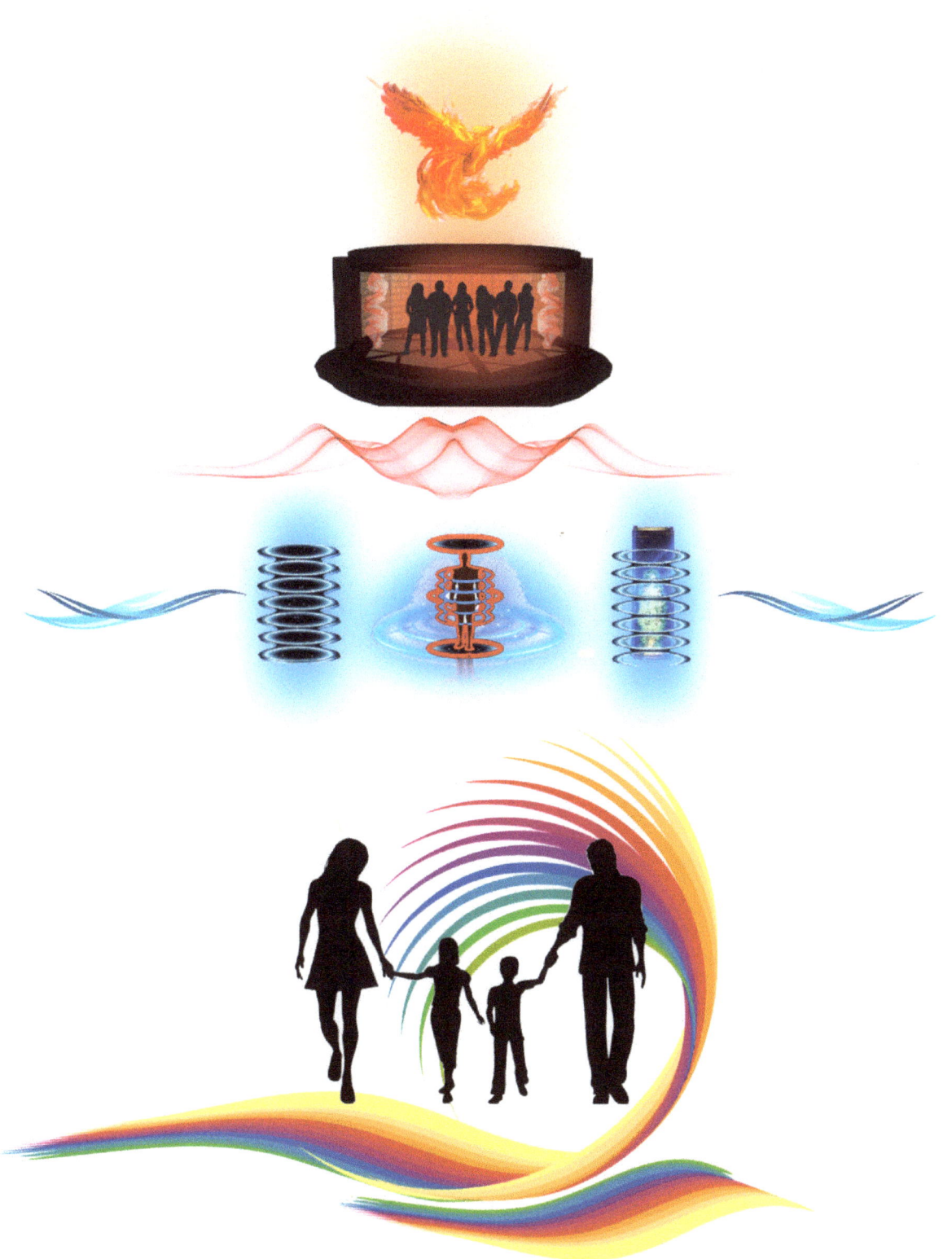

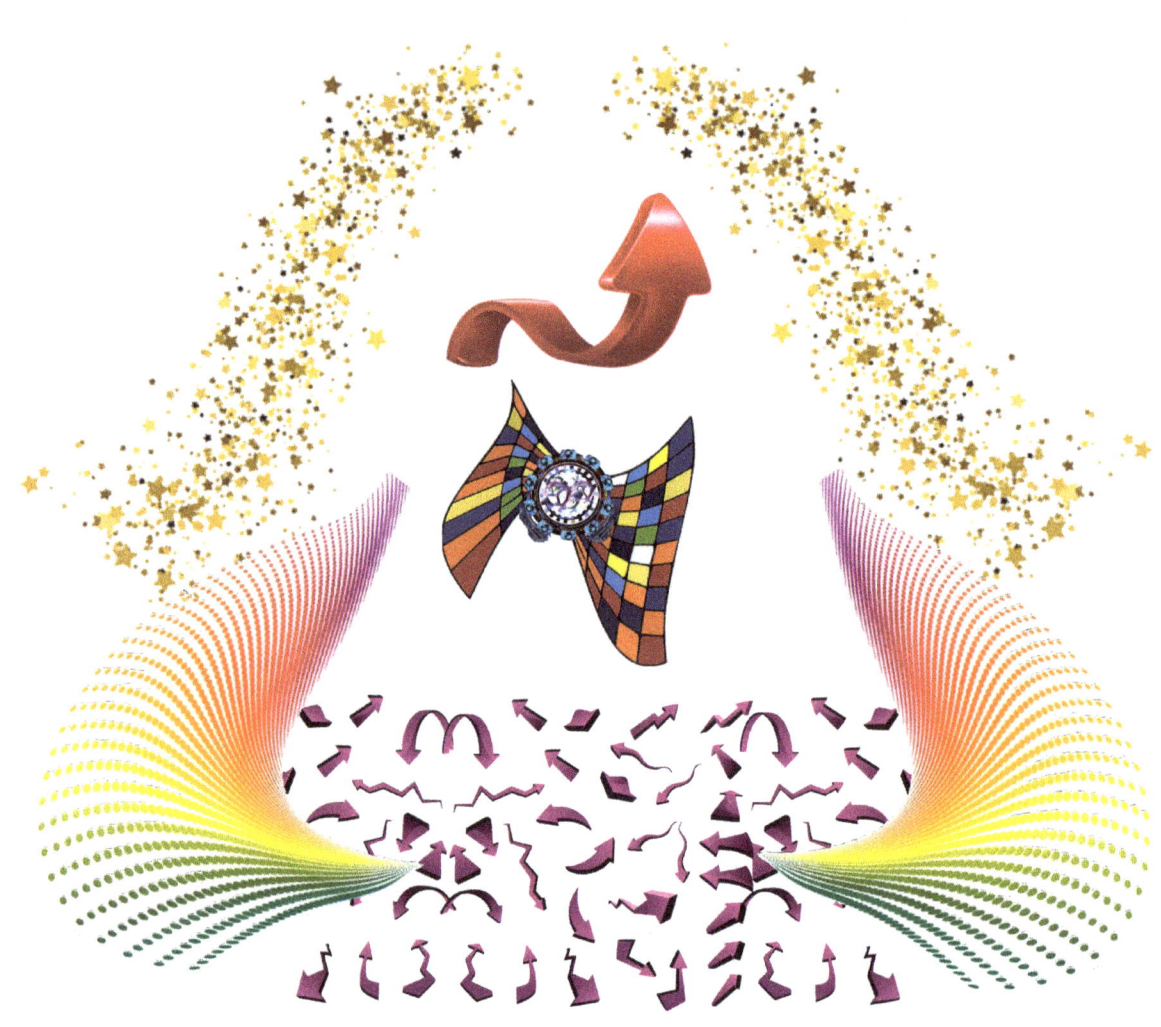

Chapter Fifteen

I'm a Doctor, Not a Doormat!
(The Holodeck Program)

Where is the boundary between reality and fantasy? In an 1889 essay entitled "The Decay of Lying", Oscar Wilde proposed that "life imitates art far more than art imitates life." In essence, he suggested that "the aim of life is to find expression, and art offers it certain beautiful forms through which it may realize that energy." In other words, "what is found in life and nature is not what is really there, but is that which artists have taught people to find there. An example posited by Wilde suggested that, although there had been fog in London for centuries, its beauty and wonder was currently noticed only because poets and painters had promoted through their art the loveliness of such effects. They did not exist until art had invented them." (Cited in Francis Charles McGrath, "Brian Friel's (post)colonial Drama" University Press, pp. 19-21).

In George Bernard Shaw's preface to "Three Plays", he wrote, "I have noticed that when a certain type of feature appears in painting and is admired as beautiful, it presently becomes common in nature; models in the picture galleries of one generation come to life as the parlor-maids and waitresses of the next." He stated that "the real world does not exist … men and women are made in the image of the fanciful creatures of their youthful fictions." (Cited in Elsie Bonita Adams "Art and Reality - Bernard Shaw and The Aesthetes." p. 76-77).

The Star Trek Universe, however, has always flown in the face of convention. In it, the Holodeck may be the quintessential example of art imitating life, and not the other way around. And it does quite a convincing job of it. The Holodeck is a device that seamlessly blends projected light, force fields, replicated matter, and other wavelengths of electromagnetic energy to create the illusion of both inanimate and animate objects. Everything within the Holodeck is solid, and crew members interact with its images and its environment in programs that are both spontaneous and predetermined. The Holodeck facilitates the introduction of locations and characters from any time in the history of the Earth or any other planet within the ship's computer data base, as well as imaginary places or beings, without having to resort to time-travel or dream sequences to create the effect.

The word "holo-graph" is from the Greek "whole" and "writing" or "drawing." In 1971, Dennis Gabor won a Nobel Prize in Physics for his work on "the holographic method" that had been initially theorized thirty years earlier. By the 22nd century, in the Star Trek Universe, the Enterprise had utilized ever more sophisticated holograms with wide latitude in regard to size and complexity. For example, the opening scene of the motion picture "Star Trek Generations" featured a holographic simulation that included an entire sailing ship bobbing on an ocean that extended to the far horizon.

In Star Trek, self-aware simulated characters raise metaphysical questions as examples of art imitating life. In at least one case ("Elementary, Dear Data"), the holographic simulation of

Professor Moriarty convincingly demanded that his sentient life be allowed to exist outside the Holodeck. On Voyager, the Emergency Medical Program and Chief Medical Officer (the "Doctor") evolved from a computer program composed of binary code, to a fully sentient character with feelings and a sense of duty, ethics, and moral responsibility. In time, the "Doctor" acquired technology from the 29th century, a "mobile holo-emitter" that allowed him free access throughout the ship, and even on away missions.

In essence, the Doctor became a digital life form capable of adapting his program. He acquired interests and hobbies, established personal relationships with humans, and fell in love (at least 4 times). As he altered his program, he evolved into a gifted and empathetic crew member, earning the respect of his shipmates, and deserving of the same indulgences as everyone else on Voyager. He developed his talents, became a playwright, an artist, and a photographer. He honed his aesthetic sense, learned to appreciate opera, and he even had offspring while trapped for three years on an alien world.

The adaptive program of the Doctor allowed him to feel guilt, as when he was forced to choose to save one life at the expense of another, during battlefield triage. He was able to daydream; he experienced fear and exhibited courage when taking over command functions when the crew had been incapacitated in battle. Like Data, he was one of a kind.

His program utilized a photonic processor that could not be replicated. Like humans, he could not be backed up, restored, or copied. In the final analysis, he was just as "alive" as any carbon-based life form on board Voyager.

He was able to activate and deactivate himself, as humans do when falling asleep or waking up. In the final episode of Voyager, he was intrigued by further potential enhancements to his program suggested by Admiral Janeway's mention of a "synaptic transceiver". We'll never know what surprises that bit of technology might have unveiled, because Janeway refused to provide more information, citing the Temporal Prime Directive. In that same episode, the Doctor finally chose the name 'Joe', after his new wife's grandfather. This prompted Lieutenant Tom Paris to remark: "It took you 33 years to come up with 'Joe'?"

During his free time on Voyager, the Doctor authored a holo-novel entitled "Photons Be Free." Its characters were based on his own experiences with his shipmates. The Federation did not view the Doctor as a person who could copyright intellectual property, but since it did recognize him as an author, he was granted the artistic license to publish his novel. It made quite a sensation throughout the Quadrant. It would be wonderful if we had literary access to "Photons Be Free." What insights might we discover if we were able to read about a hologram's feelings regarding his experiences when interacting with other sentient life-forms on a level playing field? We might yet get that chance, if extra-terrestrial beings one day choose holographic representations of themselves to make First Contact.

In Star Trek, the Holodeck was used for recreational activities, including romantic encounters, for sporting events, tactical training, and criminal investigation. Frequent users sometimes developed holo-addiction, a condition which warranted limited use. Security measures called safety protocols prevented physical injury during normal operation.

The liberal use of Holodeck programs in the Star Trek Universe raises ethical questions relating to our manipulation of the relationship between pleasure and pain, good and evil, virtue and vice, light and darkness, health and sickness, and illusion and reality, as well as between compulsion and the exercise of free-will, disadvantage and privilege, reason and excuse, asset and liability, and between expansion and restriction, action and consequence, and irresponsibility and accountability.
The real world has a plethora of moral and ethical restraints in place that guide behavior when art imitates life, but we are still given many opportunities to act out our fantasies. In the Star Trek Universe, however, one who is experiencing an unpleasant holographic episode need only say: "Computer, end program." or "Bridge" if he or she wishes to exit the program. An adage seems appropriate in both circumstances: "Be careful what you wish for. You may just get it."

If extra-terrestrial beings have advanced holographic technology such as is depicted in Star Trek, and if they make First Contact through the utilization of holo-programs, it would behoove us to treat the individuals we encounter with the same courtesy and respect we would extend to any other sentient life-form. If interstellar holo-images ask us to take them to our leaders, let's hope that we would consider it a privilege to participate in an exhilarating experience when art is given an opportunity to imitate life.

Chapter Sixteen

Travel at The Speed of Thought

In the television series "Star Trek: The Next Generation," the Traveler, whose real name was unpronounceable by humans, was a native of Tau Alpha C. He had the power to alter space, time and warp fields with his mind. He could phase in and out of time and dimension, based on his ability to focus the energy of his thoughts.

We are not there yet, but perhaps we will be, in time, no pun intended. The universe is 13.7 billion years old. On its journey to Earth, photons from the most distant stars have been traveling across space, at a constant speed of 5.8 trillion miles per year. That's 5.8 thousand billion miles, every year, for as many as 13.7 billion years. To put things in perspective, just 4,000 years ago, when starlight that is just now reaching our eyes was a mere 23,200 trillion miles from Earth, our ancestors were still fashioning tools out of stone, so we haven't been players on the universal stage for very long.

Early civilizations somewhere out there could have had well over three million times as long as we have had to figure things out. What wondrous technologies might those civilizations have developed within a time span that is almost three and a half million times as long as 4,000 years. Perhaps they have achieved Type Omega Minus civilization status. (See Chapter Five: The Unknown Possibilities of Existence).

Nevertheless, what we do have going for us is impressive, at least by our terrestrial standards (which, for all practical purposes, nullifies the "impressive" part of this sentence). Since the 1970s, the Information Age, also known as the Digital Age, Computer Age, or New Media Age, has blessed our lives with a shift from industrialization to information computerization and a knowledge-based society embedded within a global economy. Amazon and Google are good examples of corporations that have seized upon the practical applications related to the explosion in information-based technology. The transition by society to the use of technology in daily life has been dramatic, with no end in sight. But it is nowhere more profound than in the sphere of the acquisition and utilization of knowledge.

In 1900, at the dawn of the 20th century, it had taken 150 years to double all knowledge. Today, it takes only around 12 months, and soon it will be every 12 hours, according to reliable estimates. Where will we be in 10, 50, or 100 years? Maybe the sky is not the limit. (We've already been there, after all). Our brains contain several billion petabytes of information. (One petabyte is 1,000 terabytes, one terabyte is 1,000 gigabytes and one gigabyte is 1,000 megabytes). Perhaps we just need to organize the data in our brains more efficiently, so we can better utilize their resources to reach our potential. We have the Internet for comparison, which is arguably not very well organized either, as presently constituted. All the information on the World Wide Web is currently estimated to be just 5 million terabytes (TB) of which Google has indexed roughly 200 TB, or just .004% of the total. (2021).

Just consider transportation technology. Julius Caesar rode a horse, but 2,000 years later so did George Washington. As a matter of fact, the gauge of railroad track in the Nineteenth Century was initially determined by the width of a horse's rear, specifically a Roman chariot horse's rear. (Verified by Snopes). In 1592, Richard III was prompted to exclaim: "A horse! A horse! My kingdom for a horse!" (Shakespeare, Richard III, Act, 5, Scene 4). Until the invention of the steam engine, which was the driving force behind the Industrial Revolution, the horse defined the absolute speed limit at which one could travel in a horizontal direction. (1 horsepower). Henry Ford is said to have remarked: "If I had asked people what they wanted, they would have said faster horses."

In the next 100 years, there were amazing advances in transportation technology, with train travel in the year 1900 achieving a top speed of about 40 mph. In the next 50 years, transportation technology took another giant leap forward. On July 23, 1949, the De Havilland Comet became the first commercial jet aircraft, with a cruising speed of 460 mph.

In the next 20 years, transportation technology took a quantum leap forward. The Saturn 5 rocket that propelled humans to the moon weighed 6.5 million pounds and had a payload capacity of 260,000 pounds. It developed almost 8 million pounds of thrust for eight and a half minutes, achieving a speed of 7 miles per second, or 24,593 miles per hour.

In the next 10 years, the Voyager 1 spacecraft, launched by NASA on September 5, 1977, achieved the fastest heliocentric recession speed of any man-made object: 10.72 miles per second (38,592 miles per hour). Thanks to the Law of Inertia, it has continuously sustained that speed in the vacuum of space for over 44 years, (as of 2022), and has now left our solar system and entered interstellar space. (To view the real-time mission status of Voyager, go to: https://voyager.jpl.nasa.gov/mission/status/).

On another front, the Internet has made communication lightning fast, and the search bar on computer screens has become "the Sippy Cup of culture." Information technology is approaching light-speed with quantum computers on the near horizon.

And how about physics? On September 10, 2008, scientists activated the Large Hadron Collider, the world's largest and most powerful particle accelerator. It was built in collaboration with over 10,000 scientists and engineers from over 100 countries, as well as hundreds of universities and laboratories, and remains the largest and most complex experimental facility ever built. At a cost of over 7.5 billion Euros, it is arguably the costliest scientific instrument ever made, consisting of a 27-kilometer ring of superconducting magnets and accelerating structures to boost the energy of the particles along its path. Inside the accelerator, particle beams approach the speed of light, at 186,200 miles, or 299,792,458 meters, per second. The data collected from the operation of the Large Hadron Collider consists of tens of petabytes of information per year. It is being analyzed by the world's largest computing grid, housed at over 170 facilities in a worldwide network in 36 countries. As of this writing, fears of the "doomsday phenomenon" (particle collisions causing black holes) have proven to be unfounded. This is fortunate, since the creation of an artificial black hole would have resulted in the annihilation of the Earth.

Evidently, we've dodged another bullet, and it might be time to set the Doomsday Clock back a few seconds.

As we approach the cosmic speed limit, perhaps we should re-write the laws of physics to allow warp speed. Warp drive is a hypothetical faster-than-light propulsion system. A spacecraft equipped with a warp drive has been theorized to be able to travel at speeds greater than that of light by many orders of magnitude, (Warp 1 = the speed of light, and Warp 9.9 = 2,083 times the speed of light), while circumventing the relativity problem of time dilation. If warp drive technology were to create an artificial "bubble" of normal space-time surrounding the star ship, the vehicle would be able to maintain interaction with objects in normal space, while at warp speed. A theoretical solution for faster-than-light travel that models the warp drive concept was formulated by physicist Miguel Alcubierre in 1994, and NASA scientists have begun to expand upon his preliminary research to learn more about the practical applications of the technology.

But what of the enigmatic Traveler, and his ability to alter space, time, and warp fields? In the "Star Trek: The Next Generation" episode entitled "Where Silence Has Lease," Captain Jean Luc Picard says: "Considering the marvelous complexity of the universe, its clockwork perfection, its balances of this against that, such as matter, energy, gravitation, time, and dimension, I believe that our existence must go beyond Euclidean or other practical measuring systems, and that it is part of a reality beyond what we understand now."

After warp speed's wrinkles have been ironed out, will the next step be to travel at the speed of thought, as does the Traveler? Perhaps so, if mathematical equations can be formulated to embrace a time-space-thought continuum, as described by Prot, in the motion picture "KPax". He told scientists examining his claims of extra-terrestrial origin and travel at thought-speed: "Einstein never said that nothing can travel faster than the speed of light." What he did say is that something traveling slower than light is slower in all frames of reference. Therefore, it cannot accelerate to a speed above the cosmic speed limit.

Perhaps "gravitational waves that propagate at the speed of thought," a concept first suggested by the physicist Arthur Stanley Eddington, in 1922, will allow us to jump past that limit. If we can somehow come to a meeting of the minds between the abstract reality that emerges from mathematics, and the warm body of nature that we can see and feel, and touch, we just might be able to ride the exponential expansion wave of knowledge to a destination that reveals the answers we are looking for, that relate to travel at the speed of thought. If we can accomplish that, it might make interstellar communication with our galactic neighbors a practical reality.

Chapter Seventeen

The Fluidity of Time

Is it possible that there has been no contact between earthlings and extra-terrestrials because they observe the Temporal Prime Directive? Might they be living within an adjacent fold in the spacetime continuum? Do the laws of physics allow (or prevent) them from making First Contact? Does alien life mirror that which is found in the Star Trek Universe, where United Federation of Planets members were forbidden from interfering with events of any kind within pre-warp societies, and were required in all circumstances to maintain the timeline, to prevent history in the galaxy from being inexorably altered?

If Starfleet were to have any knowledge of the future, or if alternative timelines in parallel universes were found to exist, its members were forbidden to reveal any details, to avoid paradoxes that could occur within any of the following four scenarios: 1) Consistency paradoxes or grandfather paradoxes, that occur when the past is changed in any way, thus creating a contradiction. 2) Causal loops that occur when a future event is the cause of a past event, which, in turn, causes the future event. 3) A Fermi Paradox, which asks the question: "If time travel were possible, then why have we not had visitors from the future?" 4) Newcomb's paradox, which asks: If it is possible to make accurate predictions because of time travel, then do those predictions contradict or cancel out free will, for the simple reason that decisions (made through the exercise of free will) are already known to the one making the prognostications?

On June 28, 2009, Steven Hawking hosted a time-traveler's party, complete with champagne, hors d'oeuvres, and balloons. It was an experiment designed to address the Fermi Paradox. He waited – but no one from the future showed up. The invitations were only sent out four years after the party was over, which was a critical element of the experiment that avoided temporal contamination of the details surrounding the event. Those who might attend could only have found the invitation in the future, dusted off the detritus of centuries, read the summons, and then used technology fueled by the catalyst of intellectual curiosity to travel back in time and make a grand entrance at just the right moment. It was hoped that their interest would overpower their strict obedience to the philosophy of the Temporal Prime Directive.

In the invitation, Hawking provided precise GPS coordinates for the party, which was prominently held on the campus of Cambridge University, in case anyone from the future got lost while speeding back through time and space. In part, the invitation read: "You are cordially invited to a reception for Time Travelers." As he finalized preparations for the party, Hawking remarked: "I am hoping that copies of the summons, in one form or another, will survive for many thousands of years. Maybe one day someone living in the future will find the information and use a wormhole time machine to come back to my party, proving that time travel will one day be possible."

Critics argued that perhaps those in the future with the ability to travel through time would remain unaware of Hawking's invitation. But that seems unlikely, given his Twenty-first Century celebrity status. To whit, Hawking's book, "A Brief History of Time" has sold more than 10 million copies.

Hawking was inspired to organize the party by the example of Albert Einstein, who, in Hawking's words, "seemed to offer the possibility that we could warp space-time so much that we could travel through time. However, not to be a party pooper, but it is likely that warping would trigger a bolt of radiation that would destroy the spaceship, and maybe space-time itself."

Although no-one attended Hawking's party, perhaps someone from the future traveled through a wormhole to attend his memorial service, that was held 10 years later in Westminster Abbey, in London. According to the Steven Hawking Foundation, in the hope that the arc of his influence had extended to the cosmos, time travelers were cordially invited to attend that ceremony, as well. Perhaps, extra-terrestrial beings took notice of the slight deviation in the time-space continuum when Hawking's energy transitioned from mortality to the vast cosmic ocean of thought, and they came to determine the cause for themselves. We might never know if anyone responded to the invitation to travel back to pay their respects to one of the greatest theoretical physicists of the Twentieth and Twenty-first Centuries. If so, they may have gone unnoticed because they chose to blend in with the crowd, in another example of the exercise of the Temporal Prime Directive.

If your head is now spinning with these interrogatives, just remember this one simple principle: God meticulously observes the Temporal Prime Directive as it relates to His interactions with His children, even though He moves about as He wishes, through both time and space, and no one can dispute His claim that He is "the Great I Am, Alpha and Omega, the beginning and the end the same which looked upon the wide expanse of eternity, and all the seraphic hosts of heaven, before the world was made." (D&C 38:1).

It was God Who introduced the quixotic element of time into the universe, and decreed that for mortals, its arrow would only move in one forward direction, He meticulously apportioned it in discrete increments of seconds, minutes, hours, and so on, in a stroke of genius that allowed thought, feeling, creativity, and spontaneity to germinate within the fertile matrix of agency. Time gave opposition free reign to influence agency. It allowed positive consequences to follow obedience to law, and negative outcomes to be the natural consequence of disobedience. When the heavens and the Earth were organized, God reached out from His eternal vantage point and re-set the celestial clock, thereby setting in motion The Plan of Salvation. His reality, as far as we were concerned, was recalibrated to a temporal scale, whereas the reckoning beforehand had been "the Lord's time, according to the reckoning of Kolob." (Abraham 3:4).

Observance of the Temporal Prime Directive gives time the unhindered freedom to be the crucible in which we burn. It provides context and establishes order within the temporal turbulence of the cosmos, all the while maintaining the principle of free will. Jean-Luc Picard correctly observed that, although some might see time as a predator that stalks us all our lives, it is really our gentle traveling companion reminding us to cherish every moment, as it measures our approach to an event horizon beyond which lies our reunion with eternity.

In the world of everyday, time passes rather slowly, and it seems to be a Newtonian constant. He was the mathematical god of the Old Testament, and his world was static and deterministic. Some would say his view was incompatible with free will. Einstein was the mathematical god of the New Testament. His view of the world turned the discipline of physics, along with the world as we know it, upside down.

Physicists now view the world in relative terms, and time is the fluid medium within which we live, and move, and have our being. (See Acts 17:28). But, for most of us, despite our impatience with its passage, we are not yet prepared for the arrow of time to move up and down, or side to side, let alone both forward and backward. It appears that, at least in the near future, we will not be able to create a temporal paradox. Time may be a fluid medium, but its consistency is like cold molasses. Because it flows so sluggishly, "now" seems to be our moment.

All we know for sure is that each day of our lives, we are 24 hours closer to the Pleasing Bar of Christ, as long as we pattern our behavior after the Thirteenth Article of Faith. (See Moroni 10:34). "We believe in being honest, true, chaste, virtuous, and in doing good to all men. Indeed, we may say that we follow the admonition of Paul: We believe all things, we hope all things, we have endured many things, and hope to be able to endure all things. If there is anything that is virtuous, lovely, or of good report or praiseworthy, we seek after these things." There is plenty of room for moral behavior within the fluidity of time.

Life is so breathtaking that we can scarcely turn our attention to anything else. But, fortunately, our guardian that is time keeps everything from happening all at once. People talk about killing time, but it's not that we have so little; its just that we waste a good deal of it. Because lost time is never found again, is the most valuable thing we can have. We can't keep it, but we can spend it, or better yet, invest it. Every day is a bank account, and time is our currency. It's what we want most, but, concurrently, it's what we sometimes use worst. It is free, and yet it is priceless. We can't own it, although everyone possesses it, and everyone uses it, although some people try to buy it. If we lose it, however, we can never get it back. We fly on its wings, and we are its navigator. As we pass through it, we determine the direction we will take. It can be a great healer, but it is a poor beautician. None of us can escape its ravages, and it waits for no man. But it seems to stand still for a woman approaching 30.

If we bide our time, we might never find enough of it to do anything of significance. If we want it, we must take it, or even make it. But it waits for no-one. The bad news is, that time flies. It can be the wisest counselor of all, but often, there is no right time; there's just time. There may not be enough to do it right, but there will certainly be enough to do it over. Whether it's the best of times or the worst of times, it's all we've got.

Time flies like an arrow, and fruit flies like ripe bananas. Do you have a minute? Just a second. I'll sleep on it. It's all relative. It moves slowly but passes quickly. As far as time is concerned, no one is rich, and no one is poor; we've each got 24 hours to spend doing as we please. But robbing a person of time can be just as traumatizing as robbing them of money. In fact, we get "held up at the office" far more frequently than we get "held up at gunpoint." Time has a wonderful way of showing us what really matters. No measure of it is long enough, but let's start with forever. We do well when we pause in our busy lives to express gratitude to God for the supernal gift of time.

The poet may have expressed it best: "Oh, this world has more of coming and of going than I can bear. I guess it's eternity I want, where all things are, and always will be. Where I can hold my loves a little looser; where, finally, we realize time is the only thing that really dies." (Carol Lynn Pearson).

Unintentionally, Steven Hawking hinted at our relationship with the ticking clock that is calibrated to a celestial scale: "The quantum theory of gravity," he said, "has created new possibilities, in which there is no boundary to space-time … The boundary condition of the universe is that it has no boundary." ("A Brief History of Time: From the Big Bang to Black Holes", p. 136). He may as well have been describing eternity, which is unique, although it shares surprising similarities to quantum mechanics' view of time. It is an open system that cannot be defined by temporal or spatial borders.

Heavenly Father belongs to eternity, and so observes celestial laws that supersede, or do not apply to, the laws of our physical universe, including those that relate to the relativity of time. For Him, time is more than a fluid medium, it is a dynamic rushing torrent, splashing in all directions and slapping on every rock of His creations. As Paul explained to the Corinthian Saints: "The natural man receiveth not the things of the Spirit of God: for they are foolishness unto him: neither can he know them, because they are spiritually discerned." (1 Corinthians 2:14). This makes it difficult to wrap our finite minds around the mind-boggling concepts relating to the fluidity of time.

Perhaps alien explorers will be able to ride the current of time as if it were a freely flowing river. Perhaps, one day that current will cast them upon Earth's coastline, and they will wade ashore, as did the conquistadores of old, to make First Contact. But let's hope that things go better for us than it did for the Native Americans. Only time will tell.

Chapter Eighteen

The Q Continuum

In the Star Trek Universe, the Q Continuum is a higher-dimensional plane of existence inhabited by a race known as the Q. They have powers similar to those possessed by the seers of whom Ammon spoke, who knew "of things which are past, and also of things which are to come, and by them shall all things be revealed, or, rather, shall secret things be made manifest, and hidden things shall come to light, and things which are not known shall be made know by them, and also things shall be made known by them which otherwise could not be known." (Mosiah 8:17).

We do not yet possess the powers of the Q. In fact, in one episode of Star Trek, Q tells the Ferengi Grand Nagus Zek: "I worship stupidity, and you're my new god." Our poor lenses cannot discern what they see, but when our eye is single to faith, our bodies "shall be full of light." (3 Nephi 13:22). Joseph Smith confirmed the reality of that promise when he said of the revelatory process with which he was becoming increasingly familiar: "My whole body was full of light, and I could see even out at the ends of my fingers and toes." (N. B. Lundwall, "The Vision," p. 11). Every child of God potentially possesses this gift, and its expression only waits to be revealed. "If your eye be single to my glory," the Lord promised, "your whole bodies shall be filled with light, and there shall be no darkness in you; and that body which is filled with light comprehendeth all things." (D&C 88:67).

The Q are on a higher plane of existence than those of us who are trapped in normal space and time, and evidently are not subject to the same laws of physics. Their influence may be analogous to that of the Lord, of Whom the Psalmist wrote: "Fire goeth before (Him), and burneth up his enemies round about. His lightnings enlightened the world: the earth saw, and trembled. The hills melted like wax at the presence of the Lord of the whole earth." (Psalms 97:3-5).

As Mormon observed: "The dust of the earth moveth hither and thither, to the dividing asunder, at the command of our great and everlasting God. Yea, behold, at his voice, do the hills and the mountains tremble and quake. And by the power of his voice, they are broken up, and become smooth, yea, even like unto a valley. Yea, by the power of his voice, doth the whole earth shake; Yea, by the power of his voice, do the foundations rock, even to the very center. Yea, and if he say unto the earth - Move - it is moved. Yea, if he say unto the earth - Thou shalt go back, that it lengthen out the day for many hours – it is done." (Helaman 12:8-14). Thus, is manifest the awful power of His Presence. It may not be so much that God commands the Earth to tremble, but rather that it is the proximal influence of His nature that causes the foundation of our world to shake to its very core.

The Q can manipulate time, space, matter, and energy, as well. These beings are nothing more than the products of our imagination, and we know that in the beginning, in the real world, it was "the Gods (and not the Q) who organized and formed the heavens and the earth" by defining the boundaries of both the temporal and spatial universe, and the eternal world. (Abraham 4:1). The Gods manipulated time, space, matter, and energy, but they did it purely by the power of faith, They set the conditions "by which the worlds were framed, (and) all things in heaven, on the earth, or under the earth. (These) exist by reason of faith as it existed in (the mind of the Gods). Had it not been for this principle of faith, the worlds would never have been framed, neither would man have been formed of the dust. It is this principle by which Jehovah works, and through which He exercises power over all temporal as well as eternal things." (Joseph Smith, "Lectures on Faith," #1). Perhaps the writers of TNG inherited the concept of "Q" from the Prophet himself, as well as from the sacred annals of religious history.

The Q are omnipotent, omniscient, and for all practical purposes omnipresent, but in a profane way. William W. Phelps presaged the influence of the Q when, in 1856, he wrote a poem for Brigham Young, entitled "There is No End." It was later adapted to music and became a cherished L.D.S. hymn: "If you could hie to Kolob in the twinkling of an eye, and then continue onward with that same speed to fly, do you think that you could ever through all eternity find out the generation where Gods began to be? Or see the grand beginning where space did not extend? Or view the last creation where Gods and matter end? Methinks the Spirit whispers, 'No man has found pure space, nor seen the outside curtains where nothing has a place.' The works of God continue, and worlds and lives abound. Improvement and progression have one eternal round. There is no end to matter; there is no end to space; there is no end to spirit; there is no end to race. There is no end to virtue; there is no end to might; there is no end to wisdom; there is no end to light. There is no end to union; there is no end to youth; there is no end to priesthood; there is no end to truth. There is no end to glory; there is no end to love; there is no end to being; there is no death above." ("If You Could Hie to Kolob").

While the Q may imitate some of the qualities mentioned in "If You Could Hie to Kolob, and while they may seem to be invulnerable and immortal, they have not yet attained the stature of God. He was described by John the Revelator as One Whose "head and ... hairs were white like wool, as white as snow; and his eyes were as a flame of fire; and his feet like unto fine brass, as if they burned in a furnace; and his voice as the sound of many waters." (Revelation 1:14-15). Joseph Smith, who beheld God in vision, said that under His feet "was a paved work of pure gold, in color like amber ... His countenance shone above the brightness of the sun; and his voice was as the sound of the rushing of great waters." (D&C 110:2-3).

Moses had an experience with Satan that reminds us of the counterfeit nature of the power of the Q. After having experienced the glory of God, "Satan came tempting him, saying: Moses, son of man, worship me." But "Moses looked upon Satan, and said: Who art thou ... and where is thy glory, that I should worship thee? For behold, I could not look upon God, except his glory should come upon me ... but I can look upon thee in the natural man." Moses could see right through Satan's deceptions, saying "I have other things to inquire of (God), for his glory has been upon me, wherefore I can judge between him and thee. Depart hence, Satan. And now, when Moses had said these words, Satan cried with a loud voice, and ranted upon the earth." He was not about to go down quietly, without a fight. But Moses received strength, and commanded: "Depart from me, Satan, for this one God only will I worship, which is the God of glory." (Moses 1:13-23).

The Q do a fair job of mimicking the powers of God; they can even perceive events in the past, present, and future. But so, too, does the Lord our "God, even Jesus Christ," for He is "the Great I Am, Alpha and Omega, the beginning and the end, the same which looked upon the wide expanse of eternity, and all the seraphic hosts of heaven, before the world was made; The same which knoweth all things, for all things are present before (His) eyes; (He is) the same which spake and the world was made, and all things came" by Him. (D&C 38:1-3).

Both physics and the Q Continuum suggest that there are no privileged frames of reference. The galaxies are imbedded in time and attached to space, but the fabric of the universe is constantly expanding. If we ask where and when the Creation took place, the answer is everywhere and forever. The universe is temporally twisted and spatially warped, and it expands in every dimension like a balloon, which may explain why the Lord said to Moses: "As one earth shall pass away, and the heavens thereof even so shall another come, and there is no end to my works." (Moses 1:38). "For by him were all things created that are in heaven, and that are in earth, visible and invisible, whether they be thrones, or dominions, or principalities, or powers: all things were created by him." (Colossians 1:16).

The Q can manipulate the fabric of existence, weaving it into intricate patterns that boggle the mind. Perhaps the title sequence of this volume should be amended to declare: "Time and space, and even reality, are the final frontiers. These are the voyages of humanity. Its continuing mission: To explore strange new worlds, to seek out new life and new civilizations, to boldly go where no one has gone before." In 1977, NASA's John Grunsfeld said: "Someday humans will leave our cocoon in the solar system to explore beyond the orbit of Pluto. Voyager will have led the way." He was speaking of the recently launched space probes, Voyager 1 and Voyager 2, but he could have been referring to the Starship Voyager, that would take its maiden voyage in the Star Trek Franchise in 1995. In several episodes, the crew would encounter members of the Q Continuum.

The Continuum is unhindered by linear time and the three familiar dimensions of space. When Enterprise crew members are plucked from the ship by Q, we are reminded how in similar fashion, John was taken from his temporal and spatial surroundings into the presence of God. He was carried into the "depths of eternity" to the "hereafter," somehow at the same time both "here" and "after," which although vague, is about as specific as the scriptures get when referring to higher dimensions that may be similar to the Q Continuum. Thus, John described what he both saw and heard as "lightnings and thunderings and voices" from the unseen world, and as the sound of trumpets speaking to him. (Revelation 4:1 & 5). To Joseph Smith, the voice of the Great Jehovah struck a similar chord "as the sound of the rushing of great waters." (D&C 110:3).

For the Q, the arrow of time moves in all directions. Our stable temporal frame of reference allows us to live within a timeline that overlays the tapestry of our familiar three-dimensional space. It reassures us that the sun will come up tomorrow, and that there will be 24 hours in each day to address life's challenges. Without the veil insulating us from God's unrestrained, unencumbered, unreserved, and uninhibited temporal reality, which is His ever-present "now", life would be too confusing for most people. This is corroborated by the chaotic consequences of the manipulation of time by the fictional Q. In several episodes of TNG, Enterprise crew members are bewildered by sensory overload that follows the alteration of the timeline by the Q.

In the real world, when the veil that has been placed over our minds evaporates, time will be no more, and as we ease into eternity, we will become increasingly comfortable with our metamorphosis to the native and natural environment of our former home. (See D&C 84:100). "Thus it is, that we are never really at home in time," wrote Neal A. Maxwell. "Alternately, we find ourselves impatiently wishing to hasten its passage or to hold back the dawn. We can do neither, of course. Whereas the bird is at home in the air, we are clearly not at home in time, because we belong to eternity. As much as any one thing, time whispers to us that we are strangers here. If it were natural to us, why is it that we have so many clocks, and wear wristwatches? "Without the veil," he concluded, "we would lose the precious insulation so necessary for our mortal probation and maturation. Our brief mortal walk in a darkening world would lose its meaning, for one would scarcely carry the flashlight of faith at noonday and in the presence of the Light of the world." ("B.Y.U. Speeches of The Year," 1979).

The Q can move about, at noonday, in time and space, by the power of their will. Brigham Young said: "I long for the time that a point of the finger, or motion of the hand, will express every idea without utterance." When we, like the Q, are "full of the light of eternity, then the eye is not the only medium through which we see, nor the brain the only means by which we understand. When the whole body is full of the Holy Ghost, we will be able to see with as much ease, without turning our heads, as we can see before us. If you have not had that experience, you ought to have. It is not the optic nerve alone that gives the knowledge of surrounding objects to the mind. I shall yet see the time that I can converse with this people, and not speak to them. We are at present low, weak, and groveling in the dark, but we are planted here in weakness, for the purpose of exaltation." (J.D. 1:70-71).

The Q can "look down upon" our affairs from a perspective that is to us both foreign and indescribable. Brigham Young, however, made an attempt. He taught that when we die, we will go into the spiritual world, "Do the spirits go beyond the boundaries of this organized Earth?" He asked. "No, they do not. They can see us, but we cannot see them, unless our eyes are opened." ("The Contributor," 10:9, quoted in N.B. Lundwall, "The Vision," p. 55-56, see also "Discourses of Brigham Young," p. 376). This only makes sense if beings from the unseen world exist in a parallel, or perhaps a higher, spatial dimension, which may be for them like our being in a room with a one-way mirror. They, like the Q, can witness the every-day world on a whim, but to those of us trapped in the here-and-now, trying to see what lies in the other "direction" beyond the mirror's reflective surface is fruitless.

Many Star Trek episodes suggest that because we are bound by the laws and conditions of our temporal and spatial reality, all we can hope to gain by venturing into the unstable curriculum of theoretical physics is a confirmation of that which we already know, which might be only our own latent image in the mirror of experience.

The Q can move from one point in the universe to another by overriding the normal constraints of the laws of physics. At the speed of light, it would take at least 94 billion years for us, or Q, to traverse the known universe. So, in the real world, physically plodding along at light speed from point A to point B seems unlikely because those of us who have gotten the Lord's attention know that His intercession can be instantaneous, no matter in what corner of the universe He may have been about His work. Samuel was once moved to exclaim: "In my distress, I called upon the Lord, and cried to my God: and he did (immediately) hear my voice out of his temple, and my cry did enter into his ears. Then the earth shook and trembled (and even) the foundations of heaven moved and shook," as God instantly responded to the entreaty of His disciple in a powerful manifestation from His higher dimensional reality. (2 Samuel 22:7-8).

So it is, when the faithful pray to Heavenly Father. As James declared: "The effectual fervent prayer of a righteous man availeth much." (James 5:16). As we cry out to Him, God hears us with the power to immediately respond to our needs, wherever and whenever He or we may be. Existing in another dimension that is analogous to the Q Continuum might give God the ability to hear all our petitions simultaneously, without the inherent limitations of three-dimensional space, time warps notwithstanding. The omnipresent example of the Q gives God plausible credibility.

Additional insight into God's dominion may come from accounts of the creation of the Earth. Brigham Young used unusual language when referring to the earth as it existed at the time of the Fall, as well as when describing what it will be like when it receives its paradisiacal glory. He said: "When the Earth was framed and brought into existence and man was placed upon it, it was near the throne of our Father in Heaven. And when man fell ... the Earth fell into space, and took up its abode in this planetary system, and the sun became our light. This is the glory the Earth came from, and when it is (again) glorified, it will (be because it has returned) again unto the presence of the Father, and it will dwell there." (J.D. 17:143). This description of falling into space and then leaving space to return to the presence of the Father suggests the existence of adjacent spatial dimensions that are also illustrated by the characteristics of the Continuum.

The Q personify the prophecy of Micah: "The Lord cometh forth out of his place, and will come down, and tread upon the high places of the earth." (Micah 1:3). At the very least, the various scenarios and plot twists in TNG episodes involving the Q make more sense when viewed against the backdrop of religious experience. If art now imitates life, perhaps, one day, life will imitate art, and our active imaginations will provide the context for First Contact with real world extra-terrestrials who have made the journey to Earth by utilizing powers similar to those of the Continuum. (See "Are We Alone in The Universe", Chapter Fifteen: I'm a Doctor, Not a Doormat).

The Q jealously guard those powers. Heavenly Father is more benevolent than the Q, and promises to use His mighty influence to sanctify the mortal world and transform it so that it "will be made like unto crystal and will be a Urim and Thummim to the inhabitants who dwell thereon, whereby all things pertaining to an inferior kingdom, or all kingdoms of a lower order, will be made manifest to those who dwell on it." (D&C 130:9). Nevertheless, with a bit of whimsy, Q does express his feelings to Captain Picard, telling him that he is "the closest thing in this universe that (he has) to a friend." ("Deja Q").

The Q are not confined by the narrow limitations of corruptible flesh. They nurture the mind-expanding idea that when our spirits are free of incarceration within our mortal bodies, "knowledge will rush in from all quarters; it will come in like the light which flows from the sun, penetrating every part, informing the Spirit, and giving understanding concerning ten thousand things at the same time; and the mind will be capable of receiving and retaining all. Not one object at a time, but a vast multitude of objects will rush before our vision, and will be present before us, filling us in a moment with the knowledge of worlds more numerous than the sands of the seashore. Will we be able to bear it? Yes, our minds will be strengthened in proportion to the amount of information imparted. It is this tabernacle in its present condition that prevents us from a more enlarged understanding. When we burst beyond the confines of our mortal clay, we shall look not in one direction, but in every direction. This will be calculated to give us new ideas concerning the creations of God, (and) information and knowledge we never can know, as long as we dwell in this mortal tabernacle. We shall have other sources of gaining knowledge besides these inlets called senses. We will be endowed, after we leave this tabernacle, with powers and faculties of which we now have no knowledge." (Orson Pratt, J.D. 2:238-248).

The Q do not grow old. With that quality, they mirror eternity. When we move into our next apartment, time will lose all significance, and "See you later," will cease to be in our vocabulary. Time, that we too frequently viewed as a predator that stalked us all our lives, will then be fondly remembered as the companion that accompanied us on our journey through mortality, reminding us to cherish every moment. We will find that mortality itself was only one of many layers of reality, and that our perspective had been faulty because we believed it to be unique. We may be shocked to learn that it was not our natural dimension. We will come to understand why it was that we were never entirely comfortable in our mortal circumstances, that we were "strangers and pilgrims on the earth." (Hebrews 11:13). This will, in turn, explain our innate thrust always toward the future, beyond the horizon, and upward toward the stars.

We will even find that growing "old" was strictly and uniquely a quality of mortality and a brilliant mechanism designed by Heavenly Father that afforded us an opportunity to gauge the approach of our reunion with Him in eternal worlds. We will discover that because we lived in only one dimly lighted corner of reality, it was difficult for us to appreciate our potential and the power of our position, that we would one day "flourish in immortal youth, unhurt amidst the war of elements, the wreck of matter, and the crash of worlds." (Joseph Addison, "Cato," Act 5, Scene 1). From our very narrow perspective, frozen in time as it were, death seemed so distant, and its consequences so remote. Too often, we grew complacent in our indifference to the subtle message reflected in the passage of time and failed to understand its significance.

The Q are consummate multi-taskers. Hugh Nibley reasoned: "As to taking a calm and deliberate look at more than one thing at a time, that is a gift denied us at present. I cannot imagine what such a view of the world would be like, but it would be more real and correct than the one we have now. Once we can see the possibilities that lie in being able to see more than one thing at a time, the universe takes on new dimensions and God (never mind the Q) takes over." ("Zeal Without Knowledge," "Nibley on The Timely and Timeless," p. 263-264). We would then be as the Brother of Jared, who, when overshadowed by the Spirit, could look upon past, present, and future generations at once. "They all came before him, and there was not a soul that he did not behold." (Mormon 8:35).

"The heavens they are many," explained the Lord, "and they cannot be numbered unto man; but they are numbed unto me, for they are mine." (Moses 1:36). Plainly, we are dealing with two orders of mind, that of mortals, and that of God (and the Q Continuum). "For my thoughts are not your thoughts, neither are your ways my ways, saith the Lord. For as the heavens are higher than the earth, so are ... my thoughts than your thoughts." (Isaiah 55:8-9).

The Q roam the wide expanse of the galaxy, much as Federation starships do, for whom space is the Final Frontier. One particularly bold explorer, perhaps an ancestor of Jean Luc Picard, declared that he intended to go not only "farther than any man has been before me, but as far as I think it is possible for a man to go." ("Captain James Cook: Explorer, Navigator, and Pioneer," BBC). The Q might irreverently recite an even more audacious couplet: "As Q is, God once was, as God is, Q may become."

The time will come when we shall enter the spatial reality of God's Rest, when we have gained a perfect knowledge of the divinity of the work, and by our actions no longer suffer from fear, doubt, apprehension of danger, the religious turmoil of the world, or from the vagaries of men. These are the self-limiting conditions that blind us to a larger view of life. His Rest, however, is born of a settled conviction of the truth in our minds, and of confidence that His reality will one day be made manifest, but only if we master the art of celestial navigation. God's invitation is to follow Him along a trajectory that will take us across the galaxy. It is prefaced by the action verb "to come." The question is: If we come to Him, just where and when and how far will that journey take us?

After we have kept our second estate, and have had glory added upon our heads, what will we be like? What does it mean to be clothed with immortality and eternal life? Will we then more closely resemble our Father in Heaven in both His image and His likeness? If that is true, could we now be gods and goddesses in embryo? And if that is so, does it mean our genetic code itself is divine? Is it our destiny to mature to the stature of our Heavenly Parents? When we are born again, is it a process of maturation, or of generation? Can we become new creatures in Christ? Can the universe really be a machine for the making of gods? The example of the evolutionary progress of the Q, as envisioned by the writers of TNG, would suggest affirmative answers to these questions, although the scripts we would memorize, and need to follow, might require a profound attitude adjustment as they took us to new frontiers of experience.

In the final episode of Star Trek: The Next Generation, Captain Picard joins the crew's regular after-hours poker game, expressing regret that he had not done so before. The stakes are outlined, and he says: "The sky's the limit," suggesting that the Final Frontier has borders yet to be probed, opportunities to be had, adventures to be experienced, and destinies to be fulfilled.

Perhaps the interest of the Q in humanity was first kindled by the exploits of James T. Kirk, captain of the first Enterprise (NCC-1701). Kirk's middle name was Tiberius. Tiberius Julius Caesar Augustus was born 2,275 years before Kirk and was one of ancient Rome's greatest generals and tacticians. Thus, Tiberius would have been an appropriate name for the youngest graduate of Starfleet Academy to be given his own command. It may have been his bold daring that caught the attention of Q. They may have thought, as did Hamlet: "What a piece of work is (this) man, how noble in reason, how infinite in faculties, in form and moving how express and admirable, in action how like an angel, in apprehension how like a god! The beauty of the world, the paragon of animals." (Shakespeare, "Hamlet," Act 2, Scene 2).

While their powers remain intact, it would seem, at least superficially, that the Q have little need for redemption. But it would be a mistake on their part, if they were to persist in this flawed logic. In the episode entitled "Déjà Q," after he has lost his powers as a result of misconduct within the Continuum, Q is told by Captain Picard: "For all your protestations of friendship, your real reason for being here is protection." Q responds: "You're very smart, Jean-Luc. But I know human beings. They're all sopping over with compassion and forgiveness. They can't wait to absolve almost any offense. It's an inherent weakness in the breed." To which Picard retorts: "On the contrary, it is a strength.

The Q seem to exhibit a genuine interest in the development of humanity. As Q told Captain Picard: "The hall is rented, the orchestra engaged. It's now time to see if you can dance." ("Q Who"). But the Lord goes a step further, promising to prepare His children to be as He is, and giving them at least a portion of His own powers, but only after they have submitted to His instruction, demonstrated that they understand their responsibilities, and have appropriately groomed themselves. He suggests patience, "that they themselves may be prepared, and that my people may be taught more perfectly, and have experience, and know more perfectly concerning their duty, and the things which I require at their hands." (D&C 105:9-10).

The Q understand humanity's compulsion to explore the nature of existence itself. In the episode "All Good Things," Q mentions that Captain Picard was destined to consider possibilities that he had never imagined, implying that there are states that are natural to the Q but foreign to mortals. The Q become a catalyst to stir the Enterprise-D crew as a fire in their bones. Christopher Columbus recounted the similar impetus for his voyage of discovery, by simply saying: "The Holy Spirit gave me fire for the deed." Our hearts burn within us when God gives us "knowledge by His Holy Spirit, yea, by the unspeakable gift of the Holy Ghost." (D&C 121:26 & 28). Thus, did Jeremiah describe his desire to serve the Lord: "His word was in mine heart as a burning fire shut up in my bones, and I was weary with forbearing, and I could not stay." (Jeremiah 20:9).

Just as powerfully, the Spirit worked on Belshazzar's troubled conscience to the extent that his "countenance was changed, and his thoughts troubled him, so that the joints of his loins were loosed, and his knees smote one against another." (Daniel 5:6). Joseph Smith was moved to declare of his revelatory experiences: "The still small voice ... whispereth through and pierceth all things, and often times it maketh my bones to quake while it maketh manifest." (D&C 85:6).

As the process unfolds as we mature, and our unsteady steps become more disciplined, we recognize the wisdom of Hans Christian Anderson, who said: "Our lives are fairy tales waiting to be written by the finger of God." Many of the chapters in our personal journals have already been written, and we don't know how many pages remain. But we do know this: Although we cannot start over and make a new beginning, we can begin now and write a new ending. We believe God when He says: "If your eye be single to my glory, your whole bodies shall be filled with light, and there shall be no darkness in you; and that body which is filled with light comprehendeth all things." (D&C 88:67). For "that which is of God is light; and he that receiveth light, and continueth in God, receiveth more light; and that light groweth brighter and brighter until the perfect day." (D&C 50:24).

In marvelous ways, as we gain spiritual maturity, "by doing our duty, faith increases until it becomes perfect knowledge." (Heber J. Grant, C.R., 4/1934). As the seasons of our lives unfold, we learn that "life is a sheet of paper white, where each of us may write a line or two, and then comes night. Greatly begin. If thou hast time for but a line, make that sublime. Not failure, but low aim, is crime." (James Russell Lowell).

In one episode of Star Trek, Q is stripped of his powers and becomes human. ("Deja Q"). He is "the king who would be man", lamenting: "I have no powers! Q, the ordinary! Q, the miserable, Q, the desperate!" The difference between Q and our Alpha progenitors is that Adam and Eve had a longitudinal view and perspective on life that was denied the Q, who seem to have been myopic in their vision, despite their supposed omniscience. Our first parents cherished the opportunity to become mortal, and they were joyful. They felt empowered, extraordinary, and confident. "Blessed be the name of God," Adam declared, "for because of my transgression my eyes are opened, and in this life, I shall have joy, and again in the flesh I shall see God. And Eve, his wife, heard all these things and was glad, saying: Were it not for our transgression we never should have had seed, and never should have known good and evil, and the joy of our redemption, and the eternal life which God giveth unto all the obedient." (Moses 5:10-11). Due to what must have been a genetic mutation during their evolution, the Q seem to be incapable of experiencing the adrenalin rush that comes when mortals yield themselves to the power of the Atonement and become new creatures in Christ.

When Q had become mortal, he confided to Captain Picard: "Truthfully, Jean-Luc, I've been entirely preoccupied by a most frightening experience of my own. A couple of hours ago, I realized that my body was no longer functioning properly. I felt weak. I could no longer stand. The life was oozing out of me. I lost consciousness." The captain sardonically replied: "You fell asleep." Q exclaimed: "Oh, how terrifying. How can you stand it day after day?" To which the captain reassured him: "You'll get used to it."

"What other dangers await me?" Q protested. "I'm not prepared for this. I need guidance." When his powers were later restored, he exclaimed: "I'm forgiven! My brothers and sisters of the Continuum have taken me back. I'm immortal again! Omnipotent again!" Q does not realize that mortality is a brief but necessary interlude in the grand scheme of things, and that all that is necessary may be accomplished before the book is closed on this chapter in our lives. Q cannot understand that death is as much a part of life as is birth, and he has no knowledge that Adam's transgression was integral to the execution of The Plan, or that it gave him and all his posterity a power that has evidently been denied to the Q: the opportunity to be born into this world, to live, and to die.

Living in an eternal world before our births, we were able to preview the big picture and so we shouted for joy. (See Job 38:7). When it was finally our turn on earth, others smiled at our birth, while we cried. When it is our time to leave, our loved ones will cry at our departure, while we will smile in hopeful anticipation. Only then will death be seen for what it really is, "a mere comma, and not an exclamation point" in the grammar of the Gospel. (Neal A. Maxwell, "Ensign," 5/1983). It is "not extinguishing the light, but rather putting out the lamp because the dawn has come." (Ramindraneth Tagore).

The Q generally observe the principle of free will (but only if it suits their fancy). Among the shortcomings of the Continuum is the fact that it has not enthusiastically embraced the concept of opposition and does not realize that progress is dependent upon the opportunity to choose in an atmosphere that offers both positive attractions and negative enticements. It we want to be up and moving forward on the path of progress, we often must push against less desirable alternatives.

The stagnation within the Continuum that is suggested by the plot lines of several episodes of TNG reflects its ignorance of this eternal principle. However, Q did warn Captain Picard: "You judge yourselves against the pitiful adversaries you've encountered so far - the Romulans and the Klingons. They're nothing compared to what's waiting. Picard, you are about to move into areas of the galaxy containing wonders more incredible than you can possibly imagine, and terrors to freeze your soul." He continued: "It's not safe out here! It's wondrous, with treasures to satiate desires both subtle and gross. But it's not for the timid." ("Q Who?").

Q seems to know beforehand how the crew of the Enterprise will behave, and although he often orchestrates their experiences, he does begrudgingly honor the principle of free will. As he told Captain Picard: "Con permiso, Capitán. The hall is rented, the orchestra engaged. It's now time to see if you can dance." ("Q Who").

The Lord's preoccupation with our performance is equally remarkable, for He already "knoweth the thoughts of man." (Psalms 94:11). But the difference between His interest and the curiosity of the Q is that the latter look at mortality as the whole of existence, and pain, sorrow, failure, and short life as misfortunes. The vantage point of the Q Continuum may be expansive, but it lacks the capacity to develop the moral element of responsibility we call faith, that is nurtured within the learning laboratory of life built into the structure of The Plan of Salvation. God's mission statement stands transcends the agenda of the Q: His work and glory are to bring to pass our immortality and eternal life. (See Moses 1:39).

The Q are always lingering in the background, observing with interest the activities of the Enterprise crew, even though they characterized 24th Century Earth as a "dreary place" and "mind-numbingly dull." For our part, we intuitively wish we "could remember the days before our birth, and if we knew the Father before we came to earth. In quiet moments when we're all alone, we close our eyes and try to see our Heavenly home. Although we can't remember and cannot clearly see, we listen to the Spirit and so we must believe. But still we wonder, and we hope to find the answer to the question that is on our minds. Where is Heaven? Is it very far? We would like to know if it's beyond the brightest star." (Janice Kapp Perry).

The Q might have at least some appreciation of the necessity of opposition, for they allow Starship personnel to experience both the positive and negative consequences relating to their exploration of the galaxy. As Q told Jean Luc Picard after a particularly traumatic encounter with the Borg in the Delta Quadrant: "If you can't take a little bloody nose, maybe you ought to go back home and crawl under your bed. It's not safe out here. ("Q Who"). Joseph Smith similarly exhorted the Saints: "Brethren, shall we not go on in so great a cause? Go forward and not backward. Courage, brethren; and on to the victory." (D&C 128:22).

The Q recognize the fact that "the galaxy can be a dangerous place when you're on your own." ("Q-Less"). But, as Lieutenant Commander Tuvok of the Starship Voyager remarked: "We often fear what we do not understand. Our best defense is knowledge." Although Q has the ability to change circumstances by rewriting the playbook, God has the power to transform lives by encouraging our understanding of and strict obedience to His ordained Plan. Q would have us shoot the arrow blindly, and then he would impishly move the target so we would think we had scored a bullseye. God teaches us to internalize principles, practice relentlessly, and aim with the precision of doctrine, that we might hit the bullseye 100% of the time. While Q would rightly say "You miss 100% of the shots you don't take," God would encourage us to learn how to make 100% of the shots we do take.

Sometimes, the Q, like God, feel that the efforts they make on behalf of humanity are unappreciated. In "Tapestry," Q says: "I gave you something most mortals never experience – a second chance at life. And now all you do is complain." Unlike the Q, it is God's Plan that cultivates gratitude in His disciples. The Plan has the power to eclipse evil with good. It teaches that love will overpower jealousy, hatred, and prejudice. The light that is generated by its principles drives away darkness. The knowledge that it imparts banishes ignorance. The humility of its devotees overwhelms pride. Its courtesy checks rudeness.

Our appreciation of its doctrines overcomes thanklessness. The abundance of its expansive themes overshadows poverty. Our sense of well-being when we follow its precepts replaces weakness. Its simplicity supplants perplexity. Its harmony displaces discord. Our faith in its dogma controls our fear of the unknown. It builds our hope and casts out our despair. The charity we develop as we yield ourselves to Christ subdues selfishness. Our joy deposes unhappiness, sadness, dejection, and misery. Our confidence is substituted for timidity. Our certainty dethrones bewilderment. The assurance of our testimonies dislodges discouragement and even despair. With gratitude for God's Plan, our eyes are opened to the wonders of the world as though seeing them for the first time. Because of His Plan, we live in thanksgiving daily, and as a result, we are among the happiest people on earth.

Without The Plan, life can be monotonous. The boredom within the Q Continuum validates Lehi's aphorism that "there must needs be opposition in all things, else God would cease to be God. (2 Nephi 9:24). To Q3, who was overwhelmed by the dreariness of the Continuum and the apathy of its inhabitants, Q exclaims: "There's no suffering! They're all happy! Happy people! Look at them!" To which Q3 responds: "They don't dare feel sad. If only they could! It would be progress." ("Deathwish").

In the real world, "if pain and sorrow and punishment immediately followed the doing of evil, no soul would repeat a misdeed. If joy and peace and rewards were instantaneously given the doer of good, there could be no evil. All would do good, and not because of the rightness of it. There would be no test of strength, no development of character, no growth of powers, no agency, but only Satanic controls. Should all prayers be immediately answered according to our selfish desires and our limited understanding, then there would be little or no suffering, sorrow, disappointment or even death, and if these were not, there would also be an absence of joy, success, resurrection, eternal life, and Godhood." (Spencer W. Kimball, "B.Y.U. Devotional," 12/6/1955). The Q do not understand that a basic tenet of The Plan of Salvation is opposition; it is what makes our reach for the stars possible, because progress can only be measured against resistance.

The unity within the Q Continuum is only a shadow of that which is found within the congregation of the faithful. Strength, sanctuary, solidarity, safety, stability, security, and steadiness are admirable qualities of the Continuum. But it is still only a caricature of the unity of the faithful who are as one, and who can "stand independent above all other creatures beneath the celestial world." (D&C 78:14).

The Saints enjoy "the unity of the Spirit in the bond of peace." (Ephesians 4:3) Unity in purpose and conviction is one of the characteristics of the Lord's true Church, which is "of one accord, (and) of one mind." (Philippians 2:2). It is a miracle that within its organizations, all things are "done by common consent (albeit) by much prayer and faith." (D&C 26:2). Church members echo Paul, who declared: "We, being many, are one body in Christ." (Romans 12:5). No wonder that when the Lord told His disciples that they might "be one," He also warned, but "if ye are not one, ye are not mine." (D&C 38:27).

The curriculum of the Continuum stands in stark contrast to the endless possibilities afforded tho those who have matriculated into mortality's learning laboratory. Q 3 observed: "When I was a respected philosopher, I argued that the purity of the Continuum was a great thing, with infinite possibilities; only they're not so endless, after all. At the beginning of the 'New Age,' there was the exhilaration of discovery, the animated discussions of new things learned. But after a time, all had been absorbed. All had been shared. Listen to their dialogues now. They haven't spoken for millennia. There's nothing left to say!" Q 3 told Captain Janeway: "Your mission is to explore. Imagine you'd explored everything; that there's nothing left. Would you want to live forever? For us, the disease is immortality." ("Deathwish"). Q 3 described the living hell of an endless immortal vacuum without the Light of Christ, and without morality, principles, doctrines, ordinaces, covenants, principalities, power, might, and dominion. (See Ephesians 1:21).

The intellectual prowess of those inhabiting the Continuum would be equivalent to the dull, monotonous, and mind-numbing hum of a super-computer that for all its binary capabilities and permutations is incapable of experiencing real feeling, which are represented by the noblest qualities of humanity. The evolution of the Q to a plateau of omniscience and omnipotence, ithout the stabilizing influence of The Plan selectively eliminated the very things that would have guaranteed their eternal happiness.

In the episode entitled "The Q and The Grey," Q muses to Captain Janeway: "I've been single for billions of years. It was fun at first, gallivanting all over the galaxy, using my omnipotence to impress women of every species. The fact is, it's left me empty. I want someone to love me for myself. I guess what I'm saying is, I want a relationship. I just thought if you and I had a child, it would give me the kind of stability and security that I've been missing." Without realizing it, what was absent was the stability and security of family relationships that can only be found within the ordinances of the temple. (In any case, Janeway emphatically declined the invitation).

The Continuum would have done well to consider The Plan that encompasses God's ordained core curriculum leading to family exaltation, as a way out of its troubles. The Plan diagrams safe passage through the minefields of mortality, documents potential perils and possible pitfalls, charts the recommended route to refuge, maps out success strategies for abundant living, and measures progress along the pathway to perfection. The World Wide Web, that requires only computer literacy, an I.P. address with a network, and relevant hardware and software, mimics the efficiency of the Q. Its inherent danger is that its binary code is coldly logical (illogical, some would say) and utterly amoral.

Only God's Plan can order our confused and chaotic world and bless us with clarity. Only it can educate those of us who are functionally illiterate in the language of the Spirit to be fluent, that we might be mesmerized by the power of the Word. In simple terms, the elements of The Plan contain our access code and our password leading to happiness. Its curriculum is far superior to the heartless curriculum of the Continuum. The Q would have done well to heed the counsel of Jacob, who taught that when men "are learned, they think they are wise, and they hearken not unto the counsel of God, for they set it aside, supposing they know of themselves, wherefore, their wisdom is foolishness, and it profiteth them not." (2 Nephi 9:28).

The Q enjoy the "gift" of immortality, but it is eerily reminiscent of the counterfeit plan of the adversary. As Jacob taught: "Death hath passed upon all men, to fulfil the merciful plan of the great Creator", Whose omniscience, we can be sure, is greater than that of the Q. (2 Nephi 9:6). Enlightened Q 3, who finally got his priorities straight, wished: "If only I could let you see what my life is like. Oh, how I envy you mortals! The thing I want more than any other is to die!" ("Deathwish").

The Q are unschooled in the grammar of the Gospel. They may be "immortal", but at a terrible cost. They have forgotten that the three most important days of their lives are the day they were born, the day they find out why, and the day they die. They have become inured to a "resonance with realities on the other side of the veil" (Neal A. Maxwell).

Sometimes, the Q exert their powers on behalf of the crew, but they do so with a hidden agenda. In one instance, Q bestowed his upon Picard's First Officer, but Riker quickly saw through his subterfuge. ("Hide and Q"). God, on the other hand, is more benevolent than the Q, and the Latter-day work testifies to the truth of His declaration: "I have conferred upon you the keys and power of the priesthood, wherein I restore all things, and make known unto you all things." (D&C 132:45).

The existence of the Q Continuum in the Star Trek Universe underscores the necessity of The Plan of Salvation as a guiding light for an emerging species like homo sapiens that desires one day to be warp capable. It reinforces our hope that if we do encounter alien species in the real world, they will be governed by its enlightened doctrines and principles. If we meet aliens like the Q in a First Contact scenario, we can expect all hell to break loose.

Still, the Q may not be beyond redemption. Q says: "When I look at a gas nebula, all I see is a cloud of dust, but seeing the universe through your eyes has allowed me to experience wonder." ("Q-Less"). We are blessed when the eyes of our understanding have been touched by the finger of God, Who "hath given a law unto all things, by which they move in their times and their seasons; And their courses are fixed, even the courses of the heavens and the earth, which comprehend the earth and all the planets. And they give light to each other in their times and in their seasons, in their minutes, in their hours, in their days, in their weeks, in their months, in their years - all these are one year with God, but not with man. The earth rolls upon her wings, and the sun giveth his light by day, and the moon giveth her light by night, and the stars also give their light, as they roll upon their wings in their glory, in the midst of the power of God. Unto what shall I liken these kingdoms, that ye may understand? Behold, all these are kingdoms, and any man who hath seen any or the least of these hath seen God moving in his majesty and power. (D&C 88:42-47).

When the Q choose to remain hidden from view, they leave the crew of the Enterprise with few clues that would betray their presence. In contrast, God continually sends love letters to His children with His return address prominently displayed on the envelopes, so their receipt may be gratefully acknowledged and reciprocated. Truly, "earth is crammed with heaven, and every common bush with fire of God. But only those who see take off their shoes. The rest stand around picking blackberries." (Elizabeth Barrett Browning).

Perhaps the natural progression of the inhabitants of the Q Continuum was only achieved at a terrible cost, with doctrine diluted, the sacred secularized, profound truths homogenized into easily digestible forms, ennobling principles sacrificed, and expediency replacing undeviating commitment to moral standards of behavior. With sensitivity and insight, and perhaps anticipating that the Q might be past feeling, William Wordsworth penned these lines: Those like the Q "lie about us in our infancy! Shades of the prison house begin to close upon the growing boy. But he beholds the light, and whence it flows. He sees it in his joy. The youth, who daily farther from the east must travel, still is nature's priest, and by the vision splendid, is on his way attended. At length," the Q "perceives it die away, and fade into the light of common day." ("Ode: Intimations of Immortality").

The Q cannot know the exhilaration we feel when, after having made the effort to squeeze through the strait and narrow gate, our path opens into broad boulevards lined with fig trees laden with fruit, flooded by sunlight, caressed by a soothing breeze, and paved with cobblestones that glint of gold. Only then will there be no cacophony of voices assaulting us from every direction to suppress the serenity of a Gospel-centered life. Instead, we will burst free of the most glaring limitations of the Q Continuum to "slip the surly bonds of earth and dance the skies on laughter-silvered wings. With silent, lifting mind we'll tread the high untrespassed sanctity of space, put out our hands, and touch the face of God." (John G. Magee, Jr., "High Flight").

For all their powers, the Q cannot appreciate our innate desire to reach for the stars. "Humans are such commonplace little creatures," said Q. "They roam the galaxy searching for something they know not what." ("Deja Q"). The Q cannot see that our ventures are nurtured by our Father in Heaven, or that our lives are inlaid with His, as we learn to pattern ourselves after the breathtaking mosaic of His example. His power "administereth the gospel and holdeth the key to the mysteries of the kingdom, even the key of" our knowledge of His majesty. Thus, "in the ordinances" of His priesthood, "the power of godliness is manifest. And without the ordinance thereof, and the authority of the priesthood, the power of godliness is not manifest unto men in the flesh. For without this no man can see the face of God, even the Father, and live." (D&C 84:19-22). Therein lies a critical difference between the Q and Heavenly Father. Except we be quickened by the Spirit, we cannot endure God's presence, whereas our intercourse with the Q, although exciting and exotic at times, can at other times be utterly mundane and trivial. (See D&C 67:112).

Our interactions with the Q subtly and unconsciously reinforce the fact that Gospel principles carry within themselves their own witness. They need no external warrant and don't require titillation or theatrics. In contrast to the histrionic performance of the Q, Joseph Smith declared: "I (simply) teach people correct principles and let them govern themselves." (Cited by John Taylor, "Millennial Star," 13:22, p. 339). As impish as Q is in his dealings with Jean Luc Picard, to his credit, his pranks leave the captain better for having had the experience. The character development of the Q reminds us that there is hope fore all who live within the dynamic matrix of The Plan of Salvation. Neither the Q, nor we, are quite there, because perfection is a process more than it is a point.

Truth be told, were they to look at the evolution of our species objectively, the Continuum might have to grudgingly admit that the culture of our Pale Blue Dot "is richer than ever before. It is richer than that of Pericles, for it includes the Greek flowering that followed him; richer than Leonardo's, for it includes him and the Italian Renaissance; richer than Voltaire's, for it embraces all the French Enlightenment and its ecumenical dissemination. If progress is real, it is because we are born on a higher level of that pedestal which the accumulation of knowledge and art raises as the ground and support of our being. The heritage rises, and we rise in proportion as we receive it. Consider education as the transmission of our mental, moral, technical, and aesthetic heritage as fully as possible to as many as possible, for the enlargement of our understanding, control, embellishment, and enjoyment of life." (Will Durant, "The Lessons of History," p. 100-102).

In our assessment of the Q, it is apparent that those who reside in the Continuum have climbed the same cultural ladder, but somewhere along the way they missed a few rungs and lost their innocence. At first blush, the existence of the Q would seem to obviate the need for God. On closer inspection, however, the existence of the Q Continuum demands Intelligent Design, at the very least. The presence of the Q exposes our need for the irreducible qualities of faith, light, and truth to establish a reliable and stable baseline for the acquisition of knowledge. And to the chagrin of the Q, and thanks to the Light of Christ, that search must inevitably lead to testimony. As Dag Hammarskjöld observed: "The longest journey is the journey inward, for he who has chosen his destiny has started upon a quest for the source of his being."

Conversion follows testimony, as day follows night. Every time we encounter the Q, they broadcast the news that God is alive and well and is not living in hiding under an assumed name in Argentina, as some have supposed. He continues to enjoy tremendous popularity. His book is still on the best-seller list. In fact, it has enjoyed such success that He has authored additional volumes, and it is rumored that He is even now in negotiation with publishers for new book deals.

How foolish are the Q, for when they get a whiff of fame they fancy themselves as celebrities. While all the time, the character and reputation of God remains unblemished and untarnished. He, alone, deserves theatrical encores, and it is He Who, in the end, will receive standing (or kneeling) ovations from His children across the far reaches of the galaxy.

You just don't get it, do you?" said Q to Captain Picard. "The trial never ends. We wanted to see if you had the ability to expand your mind and your horizons, and, for one brief moment, you did. For one fraction of a second, you were open to options you had never considered. That is the exploration that awaits you. Not mapping stars and studying nebulae, but charting the unknown possibilities of existence." ("All Good Things"). Good for Q, but at the same time our Heavenly Father has promised to reveal "all mysteries, yea, all the hidden mysteries of (His) kingdom from days of old, and for ages to come, will (He) make known unto them the good pleasure of (His) will concerning all things pertaining to (His) kingdom." (D&C 76:7).

The Q would probably be bemused to find that their demonstration of omnipotence is evidence of the existence of God. When the Savior said: "I am come (into the world) that they might have life, and that they might have it more abundantly", He was speaking of the entire alphabet, including the letter Q. (John 10:10). He was speaking of Atonement, Baptism, Celestial glory, Deification, Exaltation, Faith unto Salvation, Grace, and so on,
all the way to Zion.

The Atonement and Resurrection of Jesus Christ were not in vain, because His disciples are still anxiously engaged, hungering and thirsting after righteousness, boldly declaring the word, and with fear and trembling working out their salvation before Him. We are carried away on the wings of the Spirit to visions of glory, we continue to smite the destroyer with the power of the word, we live life with divine fire, and we are confident that we will one day be caught up to continue our mission to explore strange new worlds, to seek out new life and new civilizations, and to boldly go where not even the Q have gone before.

Chapter Nineteen

What We Can Learn From the Q

One does not need to launch a class 1 probe to discover the differences between the shallow nature of the Q and the profound depths of awareness, and of humility, of those of us who are mere mortals but enjoy a relationship with the Spirit.

If extra-terrestrial sentient life exists in the galaxy, we have no way of determining for sure whether it is charitable in its nature, or if it is guided by the same celestial principles that motivate men and women on our Pale Blue Dot to be honest, true, chaste, and benevolent, virtuous, lovely, of good report, and praiseworthy. (See 13th Article of Faith).

But one thing we know for sure; those of us whose lives are guided by faith march to the beat of a different drummer. Our real journey to the stars is propelled by the foundation, fabric, and focus of faith. The principles that guide us are alien to the Q. They know nothing of the seeds and germination of faith, of its reach, its humility, and its knowledge, or of its power to make connections, to act, or to choose the harder right. There are weaknesses in the armor of the omniscience of the Q because they lack the testimony of faith, the faith to properly prepare, to trust, and to change hearts. They know nothing of the wisdom or the endurance of faith. We call into question the supposed omnipotence of those who inhabit the Q Continuum, because we know, as they do not, what it means to enjoy the obedience of faith, and to have the courage and moral discipline of faith.

Jeremiah asked: "Shall a man make gods unto himself, and they are no gods?" (Jeremiah 16:20). In Ephesus, "a certain man named Demetrius, a silversmith, which made silver shrines for Diana, brought no small gain unto the craftsmen; whom he called together with the workmen of like occupation, and said, Sirs, ye know that by this craft we have our wealth ... but almost throughout Asia, this Paul hath persuaded and turned away much people, saying that they be no gods, which are made with hands." (Acts 19:24-26).

The self-anointed gods of the Continuum lack the faith to believe. The nobility of faith, the divine center of faith, and the faith to be born again are alien to their nature. In contrast, the holy grail of existence for terrestrial mortals is the companionship of the Spirit. The gifts of the Holy Ghost are many; they include not only faith, but also discernment, wisdom, knowledge, administration, tongues, the interpretation of tongues, faith to be healed and to heal, prophecy, the working of miracles, the testimony of Jesus, and believing the testimony of others, to name just a few.

Heavenly Father never meant for these spiritual gifts to be handed to us on a silver platter. They are more than merchandise to be bartered in the bustling marketplaces of Ephesus and Metropolis. Perhaps, this is why the Q, as they are depicted in the Star Trek Universe, lack the admirable character traits that are linked to a performance cost. The Q may seem, at least superficially, to be omnipotent, but to understand the mystery of mysteries, to be at peace with our place in the cosmos, and to experience joy, we must read, fear, hope, and pray. We must lift the latch and force the way. (Sir Walter Scott, "A Psalm of Life"). We must expend soul sweat as we earnestly seek the best gifts. Our education, to be meaningful, must be deserved, for it must be earned with the equity of exercise line upon line, and precept upon precept. (See D&C 98:12). The Q have been wandering the cosmos for millennia; they have been tossed to and fro with every wind of doctrine, because they lack the focus of faith that The Plan of Salvation brings to the efforts of sentient life forms everywhere. (See Ephesians 4:14).

Warp drive, trans-warp drive, spore drive, subspace, coaxial space, and even travel at the speed of thought will not bring us as close to God as will simply dropping to our knees in prayer. It just might be that the Q can teach us a lesson by their poor example about accessing the ultimate power source in the universe. It is not the omega particle, or the quantum slipstream drive. (See Chapter Five – Final Thoughts). As Spencer W. Kimball suggested, the shortest and fastest route to heaven has already been plotted. The course extends from Palmyra to Paris, and not the other way around. If it ever becomes necessary to enlist the aid of an intermediary to pull off an interstellar gathering of distant cousins, it will be God Who steps forward to fill large shoes. He will send out the invitations, arrange the travel itineraries, finalize the agenda for the festivities, and make the introductions prior to and during First Contact. God, and not the Q, will be the party planner and event coordinator.

At that grand family reunion, there will be one common element that will be equally shared by all. It is the Force, an energy field with which all living things are blessed. It surrounds us and penetrates us. It is like duct tape, because it has a light side and, in accordance with the principle of opposition, a dark side. But, at its best, it does bind us all together.

Paul wrote that "we, being many, are one body in Christ." (Romans 12:5). The Spirit unites a Federation that is composed of all the children of Heavenly Father, and we live and move, and have our being because of His Charter, which is the rule of law throughout the cosmos. It is The Plan of Salvation. (See Acts 17:28). In a sense, Paul also referred to it as spiritual duct tape. "For to one is given by the Spirit the word of wisdom; to another the word of knowledge by the same Spirit; to another faith by the same Spirit; to another the gifts of healing by the same Spirit; to another the working of miracles; to another prophecy; to another discerning of spirits; to another divers kinds of tongues; to another the interpretation of tongues." (1 Corinthians 12:4, 7-10).

The Q carry out their wishes with the snap of a finger, but the Savior's disciples understand that "all things must be done in the name of Christ, whatsoever (we) do in the Spirit." (D&C 46:31). A simple thing like baptism may be the spiritual equivalent of a snap of the finger, for, in one short sentence of seventeen words, the ordinance qualifies us for membership in the Church. But there is more to the Gospel than an outward ordinance; baptism does not assure us of the spiritual transformation that is necessary for us to regain the presence of God. This comes through the baptism of fire and the Holy Ghost, one of the mysteries of God which, in this case, is the receipt of the Spirit unto sanctification. (See Moses 6:60).

Alma asked his brethren a question that would have been incomprehensible nonsense to the supposedly omniscient Q: "Have ye spiritually been born of God?" (Alma 5:14). The people in the city and land of Zarahemla had been converted to the Church, but Alma wanted to know if they had also been converted to the Savior and His Gospel. Mahatma Gandhi once said: "If a single person achieves the highest kind of love, it will be enough to neutralize the hatred of millions." Alma knew that the pure love of Christ in the hearts of his people would be a dynamic influence for good that would surpass any other motivation.

Unlike the Q, the Spirit dominates our lives only through gentle persuasion. It guides us with a still small voice, that we may know the truthfulness of all things, which is a gift that is denied the Q. Even more importantly, it quietly molds and shapes us into new creatures in Christ, something the Q cannot seem to fathom, either figuratively or literally. Foreign to the Q is the principle that it is by the power of the Spirit that our eyes are opened, to see and understand the things of God. (See D&C 76:10).

For all their self-proclaimed omniscience, it is we, and not the Q, who find our powers expanding when we experience the glittering facets of the life of the Spirit. We use our careful preparation and training as a springboard to disciplined, controlled procedure, and we are receptive to flashes of insight. The principles of The Plan comprise our operations manual, and our obedience to its instructions sets us free to be creative and sets us creative to enjoy greater freedom. As we learn to respond to the guidance of the Spirit, we become more familiar with its gifts. It is the perfect law of liberty.

Each of us is given "a gift by the Spirit of God." (D&C 46:11). When His image has been engraven upon our countenances, we "shall ascend into the hill of the Lord (and) stand in his holy place" to partake more fully of the Divine Nature. We will have clean hands and pure hearts, and we will not lift up our souls unto vanity, nor sware deceitfully. (See Psalms 24:4-5). We will model our behavior after our Father, and not after the poor example of the Q.

Sometimes, we must fast and pray to receive strong, independent testimonies, and to gain a witness of the Spirit. Whatever the price, however, until it has been paid we will not be able to comprehend its language. If we have never made the journey to Christ, if we have not dutifully trudged along the path leading to the tree of life, and if we have not worked tirelessly to then harvest the delicious fruit of that tree, we will never be able to receive "the things of the Spirit of God, for they (will be) foolishness unto (us), neither can (we) know them, because they are spiritually discerned." (1 Corinthians 2:14-15). The Q have never learned the fundamental lesson of life that faith precedes the miracle of the receipt of spiritual gifts. Faith is the most powerful force in the cosmos; by it, as many as 1.6 billion new worlds are created every day. (See Chapter Eleven: The Universe is a Star Nursery). And that is an accomplishment that is beyond the capabilities of even the Q.

As we grow in spiritual stature. We begin to see more clearly the path that we must follow. We prayerfully consider the counsel of the Savior to "be perfect, even as (He and our) Father who is in heaven (are) perfect." (3 Nephi 12:48, see Matthew 5:48). We glory in the possibility that when we obey God and endure to the end in righteousness, we will inherit His character, that is personified by His image in our countenances and His likeness in our nature. (See Genesis 1:26). It is our noble birthright to become more like Him during our mortal walk. It is interesting that, although Starfleet was aware of the existence of the Q early on, there was no inclination to emulate them. There was no course at Starfleet Academy that taught cadets to imitate the behavioral subtleties of the Q. The Continuum was almost universally considered nothing more than a curious novelty and a cosmic anomaly.

God, on the other hand, offers His matchless grace. Only He has the power to sanctify us, "through the shedding of the blood of Christ ... (that we) become holy, without spot." (Moroni 10:33). This is the essence of the Gospel. If we open our hearts to the principles of The Plan of Salvation, we can become as lambs without spot or blemish. (See. 1 Peter 1:19, & Hebrews 9:14).

Isaiah invited those in the arid regions of space, "everyone that thirsteth, (to) come ... to the waters, and he that hath no money (to) come ... and eat; yea, come, buy wine and milk without money and without price." (Isaiah 55:1). We cannot purchase spiritual gifts with gold-pressed latinum or the profane baubles and treasures of the earth. When we see a need, we cannot throw obscene amounts of money at it, and expect to solve our problems.

Perhaps, this is why in their efforts to obtain sacred records, Lehi's sons were stripped by Laban of all their precious things. The task was to be accomplished in the Lord's way, by the power of His mighty arm. (See 1 Nephi 3).

Think about what an influence the Q might have been, if they had been able to utilize the gift of faith to motivate Starfleet personnel to action, to fulfil the mandate of the Federation's Charter. Even more, they could have helped to prepare the way for all 151 member civilizations of the Federation to be "partakers of the heavenly gift." (Ether 12:8). Alas, the Q proved to be no match for the Savior, Whose nurturing influence at another "First Contact" of sorts prompted the two disciples on the Road to Emmaus, to declare: "Did not our heart burn within us, while he talked with us by the way, and while he opened to us the scriptures?" (Luke 24:32).

Those discerning disciples demanded no visible proof of life among the stars; rather, they enthusiastically acknowledged the hand of the Lord as an influence in their lives. The Spirit provided a noble way to discern the way, the truth, and the life. (See John 14:6). To increase faith, the Lord did not indulge the prurient interest of those who only want theological titillation to satisfy their adulterous curiosity. He did not pander to the base instinct of a fallen and depraved nature.

The gift of wisdom, which includes the ability to discern the difference between knowledge and intelligence, is not capriciously bestowed, as the Q would have it. Instead, it is given to those who press forward toward the tree of life with complete dedication, along the way feasting upon the words of Christ and thereby receiving physical and spiritual strength and nourishment. Those who endure to the end with continuing responsibility and accountability will be given the promise of hidden treasures of knowledge, leading to intelligence, or light and truth. The omniscience of the Q, it would seem, has only been achieved at the cost of saving faith, and as a result their vessels of oil are never replenished, and they will be empty at the day of reckoning when First Contact is made with the Spirit.

Paul declared that "God hath chosen the foolish things of the world to confound the wise; and (He) hath chosen the weak things of the world to confound the things which are mighty." (1 Corinthians 1:27). The Lord instructed Joseph Smith to teach the doctrine of the kingdom to the end that all might be edified in Christ. (See D&C 88:77). He also instructed that we should seek not for riches, but for wisdom. To understand spiritual things, we must have insight, intuition, discernment, inspiration, and revelation, or guidance from the Holy Ghost. We must listen to the still small voice. The mysteries of God are those truths that can only be known by the ministrations of the Spirit. These are principles that completely elude the prodigious intellectual capacity of the Q. When the obedient faithful "hunger and thirst after righteousness," however, the doctrine of the priesthood will distill upon their souls as the dews from heaven, because the Holy Ghost will be their constant companion. (See D&C 121:45-46).

Our hearts are stronger than those of the Q, because they are pliable. They are open and receptive to the greater portion of the word, until it is given unto (us) to know the mysteries of God until (we) know them in full." (Alma 12:10). The members of the Continuum have never learned that to have access to the spiritual gift of knowledge of the mysteries of God, and to understand "the peaceable things of the kingdom", we only need to be teachable. (D&C 39:6).

The Achilles Heel of the Q is that when they harden their hearts to the truth, they are "given the lesser portion of the word until they know nothing concerning (God's) mysteries, and then they (will be) taken captive by the devil and (be) led by his will down to destruction." (Alma 12:11).

The terrible thing about anyone, Q or otherwise, hardening their hearts is that their understanding is withheld. Skeptics and cynics are more vulnerable to the devil's influence. The guidance of the Spirit is withdrawn from doubters, leaving them alone to grope in darkness. On their own, they cannot claw their way out of the pit they have dug for themselves, and in the absence of atonement, they have no claim on blessings, prosperity, or preservation. Having eyes, they see not, and having ears, they hear not. (See Matthew 13:15). They are as the blind leading the blind. (See Matthew 15:14). They disqualify themselves by unworthiness of First Contact.

It is plain to see that it is the habit of those who feel uncomfortable in proximity to spiritual experiences to withdraw to lifestyles devoid of such associations. Thus begins a downward spiral that can only gain momentum as self-defeating behaviors become entrenched. Even worse, those "that doeth this, the same cometh out in open rebellion against God." (Mosiah 2:37). For all their powers, the Q suffer spiritual death, or life that is devoid of light and truth. The Q exist in the cold, unforgiving vacuum of space, where dark matter prevails.

Moroni wrote that to some it is given that they "may teach the word of knowledge." (Moroni 10:10). Joseph Smith taught that "it is impossible for us to be saved in ignorance" of the saving principles of the Gospel." (D&C 131:6). We may have mastered quantum mechanics, exo-biology, and astrophysics, but we must turn our attention to, and embrace, the principles and doctrines of the Gospel. We must have knowledge of them and of our Heavenly Father, and then apply that knowledge in practice. For Jesus taught: "This is life eternal, that they might know thee the only true God, and Jesus Christ, whom thou hast sent." (John 17:3). The knowledge that we have of them is a gift of the Spirit. Unless the Q first humble themselves, they are unlikely to receive that gift or secure a relationship with Deity. They will continue to live the lie that their powers are heaven-sent, and their omniscience is more than a façade.

The Q know nothing of the ordinances that bridge the gulf between heaven and earth. If they did, they would understand that these covenants of action illustrate that the requirements for obtaining salvation are the same for all of Heavenly Fathers' children, no matter with what letter of the alphabet they may identify, or where in the cosmos they may wander. (See Acts 10:34, & 1 Nephi 17:35).

The Q have no idea that they operate on borrowed light. They roam the galaxy looking for species who might amuse them. They seek form without substance and enjoy the relative ease of putting forth minimal effort that makes no demands for personal sacrifice. One would hope that somewhere the galaxy they would stumble upon a universally applicable principle equally valid in all quadrants that allows no room for variation. It is not the omega particle, but the Spirit that powers the Church of Christ.

Perhaps the Q should try the universal translator that we call the Spirit when they are confronted by unsearchable mysteries. On our own world, with relative ease and without the aid of technology, even the humble missionaries enjoy the gift of tongues and of the interpretation of tongues. In our lore, we remember the Nephite little children whom Jesus blessed. These received an endowment of spiritual power, for the Savior "did loose their tongues, and they did speak unto their fathers great and marvelous things, even greater than he had revealed unto the people." (3 Nephi 26:14). The multitude "both saw and heard these children; yea, even babes did open their mouths and utter marvelous things; and the things which they did utter were forbidden that there should not any man write them." (3 Nephi 26:16).

We exercise the gift of the interpretation of tongues so that we might hear the words of eternal life with clear and unambiguous understanding. These words "are not of men nor of man, but of me," said the Lord, "wherefore, you shall testify they are of me and not of man. For it is my voice which speaketh them unto you; for they are given by my Spirit unto you, and by my power you can read them one to another; and save it were by my power you could not have them. Wherefore, you can testify that you have heard my voice and know my words." (D&C 18:34-36).

The Q can repair and restore the biological tissues of living organisms, but they do it in a profane way, without the gift of faith to be healed. Their robotic surgery may be performed with precision, but it lacks any semblance of bedside manner, and it is devoid of empathy and compassion. They do the right thing for the wrong reasons, and they do it without authority unto the damnation of their souls.

During the Savior's post-mortal ministry among the Nephites, He "healed all their sick, and their lame, and opened the eyes of their blind and unstopped the ears of the deaf." (3 Nephi 26:15). His efforts to bless the people were catalyzed by their great faith in Him. The Savior has no need for lukewarm converts; He would rather that we were hot or cold, but He embraces those whose commitment is profound. (See Revelation 3:15). Only then, can He truly bless our lives. After entering the Fold, our faith springs into action, and we see and hear "unspeakable things, which are not lawful to be written." (3 Nephi 26:18).

We remember the counsel of the Master, Who said: "Ye know the things that ye must do in my church; for the works which ye have seen me do that shall ye also do." (3 Nephi 27:21). Without realizing it, the Children of Christ who exercise the diligence of faith to be healed undergo a metamorphosis into a Zion society, for they are pure in heart. (See D&C 97:21).

We remember the gift of faith to restore physical and spiritual health when we read in Malachi: "Unto you that fear my name shall the Sun of righteousness arise with healing in his wings." (Malachi 4:2). There is a multi-dimensional aspect of faith to heal. Only six generations after Adam, "Enoch looked upon the earth; and he heard a voice from the bowels thereof, saying: Wo, wo is me, the mother of men; I am pained, I am weary, because of the wickedness of my children. When shall I rest, and be cleansed from the filthiness which is gone forth out of me? When will my Creator sanctify me, that I may rest, and righteousness for a season abide upon my face?" (Moses 7:48). When we ponder the significance of the terrible pollutions that mar the face of the earth, and we consider the physical and spiritual cleansing that will be required of it before the Kingdom of God can return to accept it, the concept of "healing" moves to a new level. We are comforted by the assurance that, although the Q can move mountains, and in the annals of Star Trek was, in fact, capable of moving entire star systems, the Lord can move both heaven and earth. (See Matthew 28:18).

At the millennial day, all shall raise their voices and sing, declaring that "the earth hath travailed and brought forth her strength," as a mother who has borne a new child, "and the heavens have smiled upon her," for she is pure and delightsome. "And she is clothed with the glory of her God," adorned in the strength of His priesthood. "For he stands in the midst of his people (with) glory, and honor, and power, and might. For he is full of mercy, justice, grace and truth, and peace." (D&C 84:101-102). Intention is a key to the healing power, and by most accounts, the Q fail the test, because their motivation is flawed, and they tend to be self-serving in their actions. Heavenly Father, on the other hand, has great compassion for His people. (See 3 Nephi 17:6).

The Apostle Peter indicated that true Saints "have obtained like precious faith with us through the righteousness of God and our Saviour Jesus Christ. According as his divine power hath given unto us all things that pertain unto life and godliness, through the knowledge of him that hath called us to glory and virtue; whereby are given unto us exceeding great and precious promises: That by these ye might be partakers of the divine nature." (2 Peter 1:1, 3 & 4). Peter was not describing the Q, but those who have experienced the power of God, understand His character, and know that the whisperings of the Spirit will never lead them astray.

Ammon taught King Limhi that "a seer is a revelator and a prophet also; and a gift which is greater can no man have, except he should possess the power of God, which no man can." (Mosiah 8:26). The gift of prophecy is "the testimony of Jesus." (Revelation 19:10). It follows that any sentient carbon-based life form who has received the spiritual gift of a testimony is a prophet, since it can only be received by revelation from the Holy Ghost, and since it consists of the words that holy men speak when they are moved upon by the Spirit. On this point, the jury is still out on the Q, and for the time being, we will give them the benefit of the doubt. They seem to exercise prophetic omniscience, but whether they are manipulating the Temporal Prime Directive for their own amusement or to their own advantage is open to debate.

Rather than aspiring to be as the Q, we should praise and glorify God. We do that by not denying the revelations of the Lord, or by protesting that He "no longer worketh by revelation, or by prophecy, or by gifts, or by tongues, or by healings, or by the power of the Holy Ghost!" (3 Nephi 29:6). In truth, we do not need the Q, because of our firm and abiding testimony "that revelation continues, and that the vaults and files of the Church contain these revelations which come month to month, and day to day." (President Spencer W. Kimball, C.R., April 1977). Our vessels are full of oil.

As we move along the Highways and Byways of Life, we remain "at work with our hands to the plough and our faces to the future," as Sir William Mulock reflected. "The shadows of evening lengthen about us, but morning is in our hearts. Ours is this: The castle of enchantment is not yet behind us, but is before us still, and daily we catch glimpses of its battlements and towers. The best of life is always further on. The real lure is hidden from our eyes, somewhere behind the hills of time." (A Complimentary Luncheon to The Right Honourable Sir William Mulock, at the Empire Club of Canada, 2/13/1930).

The gift of the working of miracles is a manifestation of the power of God, and at the end of the day, Q "receiveth not the things of the Spirit of God: for they are foolishness unto him: neither can he know them, because they are spiritually discerned." (1 Corinthians 2:14).

Those who enjoy the gift of a testimony of Jesus Christ unflinchingly bear witness that He was "the Son, the Only Begotten of the Father, full of grace, and mercy, and truth." (Alma 5:48). Either Jesus "was, and is, the Son of God, or else a madman or something worse. But don't let us come with any patronizing nonsense about His (only) being a great human teacher" or simply another member of the Q Continuum, come to Earth in disguise, "as the secular apologists would have us believe." (C.S. Lewis).

Alma taught that preaching the Gospel with power and authority is the responsibility of those who bear the priesthood of God. "This is the order after which I am called," he explained, "yea, to preach unto my beloved brethren, yea, and everyone that dwelleth in the land." (Alma 5:49). He did not make any distinction between member and non-member, "black and white, bond and free, male and female (or) Jew and Gentile," or even between the 26 letters of the alphabet. (2 Nephi 26:33). All were alike unto God.

Rather, Alma felt that it was his duty to bring the gift of the Gospel to all, for his message was the same, that all must repent and be born again, not only on our Pale Blue Dot, but on every other sphere created by God in the habitable zone of its star system. In the TNG episode "Deja Q" when he is stripped of his powers by the Continuum, Q tells Picard: "I could have chosen to exist as a Markoffian sea lizard, or a Belzoidian flea. Anything I wished, as long as it was mortal. And since I had only a fraction of a second to mull it over, I chose this, and asked them to bring me here. Q the miserable. Q the desperate. What must I do to convince you people?" According to Alma, the answer is that Q and every other sentient being in the universe must repent and be born again.

Some of those who do, have many of the qualities of the Q, but they are much more than that. Some of them have "all those gifts, that there may be a head, in order that every member may be profited thereby." (D&C 46:29). When we generate faith in the atoning power of Christ and possess the resolve to do whatever is necessary to activate the dynamic energy of His sacrifice in our own lives, we will share with others the gifts of the Spirit. We will have the overarching desire to live in accordance with His will. We will look forward with an eye of faith, with an eternal perspective. We will not only believe in Christ, but we will believe Christ, when He says that we can be, not just stellar, or galactic, or universal, or multi-versal, but celestial in our nature. The manifestation of spiritual gifts will dramatically presage our approaching stardom. Those who aren't valiant in the testimony of Jesus, who don't stand for something, will fall by the wayside, and be swallowed up in a black hole of oblivion, where not even light can escape. When individuals, including the Q, don't know where they are going, they will end up somewhere else, and in their indifference they probably won't even care that they made the trip.

A vital distinction between ourselves and the Q is that our gifts are given "for the benefit of the children of God." (D&C 46:26). In truth, the Q live a telestial existence. The Light of Christ will always attempt to influence their behavior, so even if they turn to the shadows, they will find it difficult to justify their actions before God, in the face of the many signs that have testified of Him. "Any man," including the Q, "who hath seen any or the least of these (wonders) hath seen God moving in his majesty and power." (D&C 88:47). "Earth (itself) is crammed with heaven, and every common bush with fire of God. But only those who see take off their shoes." The Q, however, "stand around picking blackberries." (Elizabeth Barrett Browning).

Ours is the Age of Inspiration and of gifts of the Spirit. With prophetic foresight, Joseph Smith promised: "God shall give unto (us) knowledge by His Holy Spirit, yea, by the unspeakable gift of the Holy Ghost, that has not been revealed since the world was until now. (This is a time when) nothing shall be withheld. All thrones and dominions, principalities, and powers, shall be revealed. And, also, if there be bounds set to the heavens or to the seas, or to the dry land, or to the sun, moon, or stars, (all this) shall be revealed in the days of the dispensation of the fulness of times." (D&C 121:26-31). The promise within this revelation far surpasses any wonders that might have revealed to the Q, to Starfleet captains, or to any of the rest of us who see only blackberries, and not the fire of God, in common bushes.

The Q lack the benevolence of our Father in Heaven. From Him, we "receive the gifts of sensory delight, of fragrance, sound, form, and color. Ours is the realm of human associations, of gratitude, loyalty, and appreciation, of selflessness, helpfulness, and forgiveness, of friendship, love, and compassion. It is the realm of human growth and transcendence and of truth discovered and accepted, of beauty created and enjoyed, and of goodness deepened and made manifest in life. None of us are strangers to these realms of spirit. We have sensed the world about us, smelled its fragrance, heard its sounds, and glimpsed its form and colors. We have warmed our souls in the glow of human associations; we have had our moments of selflessness, gratitude, love, and forgiveness. We have felt an upward reach within us when made suddenly aware of a truth, a beauty, a goodness above and beyond our own attainment." (P.A. Christensen, "A Land Unpromised and Unearned," "B.Y.U. Studies", Autumn, 1975).

Whisperings of the Spirit confirm that there is more to life than outward observances, obedience, and covenants. Q would demean these things, in order to understand and control them. In contrast, our Heavenly Father relies upon our spiritual enlightenment to awaken our discovery and expand our awareness of undreamed vistas of otherwise inaccessible experience. This is the element of wonder that escapes those who inhabit the Q Continuum.

The Apostle Paul testified that "eye hath not seen, nor ear heard, neither have entered into the heart of man, the things which God hath prepared for them that love him." (1 Corinthians 2:9). Nephi declared: " No tongue can speak, neither can there be written by any man, neither can the hearts of men conceive so great and marvelous things as we both saw and heard." (3 Nephi 17:7). Joseph Fielding Smith, Jr., described the spiritual "impressions on the soul that come from the Holy Ghost (as) far more significant than a vision. It is where Spirit speaks to spirit, and the imprint upon the soul is far more difficult to erase."

The second Book of Kings alludes to the thin line separating the temporal and the celestial worlds: "And when the servant of the man of God was risen early, and gone forth, behold, an host compassed the city both with horses and chariots. And his servant said unto him, Alas, my master, how shall we do? And he answered, Fear not: for they that be with us are more than they that be with them. And Elisha prayed, and said, Lord, I pray thee, open his eyes, that he may see. And the Lord opened the eyes of the young man; and he saw: and, behold, the mountain was full of horses and chariots of fire round about Elisha." (2 Kings 6:15-17).

"When you lay down this tabernacle, where are you going?" asked Brigham Young. "Into the spirit world," he continued. "Where is the spirit world? It is right here. Do the spirits go beyond the boundaries of this organized Earth? No, they do not. They can see us, but we cannot see them, unless our eyes are opened." ("The Vision," p. 55-56).

A reminiscence by a friend of Joseph Smith reflects the gossamer fabric of the veil separating the world we know from the world of spirits. "I am getting tired and would like to go to my rest, said Joseph. His words and tone (both) thrilled and shocked me. Like an arrow, (they) pierced my hopes that he would long remain with us, and I said, as with a heart full of tears: Oh Joseph, what could we, as a people, do without you and what would become of the great Latter-day work if you should leave us? He saw and was touched by my emotions, and in reply he said, Benjamin, I would not be far away from you, and if on the other side of the veil, I would still be working with you, and with a power greatly increased, to roll on this kingdom." ("The Vision," p. 140-141).

This is our challenge. It is not to possess the omniscience and omnipotence of the Q, because we do not need it, nor do we want it. Our quest is to learn to understand with fluency the language of the Spirit, which is available to all of God's children. This is the exploration that awaits us. "Not mapping stars and studying nebulae, but charting the unknown possibilities of existence." (Q, of all people, in "All Good Things").

The problem is that we "tend to fill space, as if what we have, what we are, is not enough. Being affluent, we strangle ourselves with what we can buy, and with things whose opacity obstructs our ability to see what is really there." (Gretel Erlich, "The Atlantic Magazine"). Our exercise of spiritual gifts is the antidote for poisonous telestial tendencies that smother the expression of celestial sureties.

And so, we "give thanks unto God in the Spirit for whatsoever blessing (we) are blessed with. And (we) practice virtue and holiness before (God) continually." (D&C 46:32-33). The gifts of the Spirit enlarge the capacity of our hearts, and bless us with the consciousness of victory over ourselves, and of communion with the infinite, which is far greater than any interaction we could possibly have with the Q.

God leaves us richer today than we were yesterday, and especially if we have laughed often, given something, forgiven even more, made a new friend, changed stumbling blocks into stepping-stones, if we have thought more in terms of others rather than ourselves, or if we have managed to be cheerful even when we were weary. We are richer today than we were yesterday if we have used the gifts of the Spirit to bless the lives of others. We receive the image of Christ in our countenances and experience a mighty change in our hearts. "This feeling is indescribable, but it is real. Happy is the person who has truly sensed the uplifting, transforming power that comes from this nearness to the Savior, this kinship to the Living Christ." (David O. McKay).

We develop our gifts, remembering that whatsoever is good cometh from God," and not the Q. (Alma 5:40). The gifts of the Spirit are given so that we may have a fortifying influence to combat evil in the world around us. Unlike the Q, there have been woven into our sinews threads of moral responsibility that we describe as faith. "We have no excuse to err in our knowledge and understanding of right and wrong. By inquiring of the Lord and listening to the voice of His Spirit, and having a willingness to be guided thereby, we will always find ourselves on the Lord's side of every issue and be strengthened to hold fast to that which is good." (Delbert Stapeley, C.R., 4/1965).

When we take His side, the windows of heaven will be opened unto us, and the intangible will "not be seduced by evil spirits, or doctrines of devils, or the commandments of men," or by those who are simply misguided, as are the Q. (D&C 46:7). Our mission statement transcends that of Starfleet which is to defend its home from aggression, to maintain peace and freedom within Federation borders, to explore strange new worlds, to seek out new life, and new civilizations, and to boldly go where no one has gone before. Instead, our quest is summarized in the last verses of The Book of Mormon: We determine to "come unto Christ, and lay hold upon every good gift, and touch not the evil gift, nor the unclean thing. Yea, (as we) come unto Christ, (we know that we will be) perfected in him, and (we will deny ourselves) all ungodliness." (Moroni 10:30 & 32).

To accomplish our mission, "that (we) may not be deceived, (we) seek ... the best gifts, always remembering for what they are given." (D&C 46:8). God will empower our ministry, and work miracles as "long as time shall last, or the earth shall stand, or there shall be one man upon the face thereof to be saved." But "if these things have ceased wo be unto the children of men, for it is because of unbelief, and all is vain." (Words of Mormon 1:36-37). The Last Days mirror those of Mormon, who wrote that "there were sorceries, and witchcrafts, and magics, and the power of the evil one was wrought upon all the face of the land" because of the lack of faith among the people. (Mormon 1:19).

When the gifts of the Spirit are absent, we must declare, as did Mormon, "faith (has) ceased also; and awful is the state of man, for they are as though there had been no redemption made." (Words of Mormon 1:38). The mercurial Q come and go as they please with frustrating unpredictability. We, on the other hand, stand in holy places and we are steadfast and immovable. (See D&C 87:8). We understand things more clearly than do the Q. We do not procrastinate the day of our repentance or wait to develop saving faith until we are spiritually dead to the Light of Christ. We are not subject to evil spirits as are those we can no longer make the vital distinctions between good and evil or light and darkness. We do not risk having the Spirit of the Lord withdraw from us, because we have voluntarily surrendered to the devil our agency to act independently. (See Alma 34:35).

Mormon was unaware of extra-terrestrial life forms such as the Q, but in his exhortation to his brethren to rise to the occasion, he may as well have been thinking of them, too. He said: "I judge better things of you." (Moroni 7:39). He was like wise old Tevya in The Fiddler on The Roof, who told his daughters: "In Anatevka, God knows who you are, and what you may become." Tevya knew that when our spiritual sensitivities are dulled, we are tempted to risk all for the thrill of the moment. Like Mormon, Tevya was unfamiliar with the Q, but he was wise enough to warn his daughters that there are those in the world who will always have some magic elixir to tickle our senses, but they will never be able to instill within us an appreciation of the wonders of eternity.

At times, the Q create and promote the illusion of spiritual manifestations, counterfeit though they may be. So it is, in the real world. Some demand signs from the ministers of the Lord as proof of their authority. (See John 2:18, 6:30). Others who have adulterous hearts seek signs from those masquerading as the Q for the satisfaction of their desires, although they require greater and greater intensities of validation for the same level of gratification. (See Matthew 12:39). The Q might be able to just walk away, but signs thus given leave the wicked with responsibility for what happens next. Because consequence follows action, signs establish accountability. (See D&C 63:7 & 11).

Mormon wrote that the people who lived in Zarahemla just before the birth of the Savior attributed the signs that had been given to "the power of the devil." They did this" to lead away and deceive the hearts of the people." (3 Nephi 2:2). This is a classic illustration of the subtle craftiness of Satan, who would even bring attention to himself if it would somehow serve his own evil purpose. The Q, in these situations, might play right into the hands of the adversary.

In reality, the devil is often the source of our rationalizations: "And thus did Satan get possession of the hearts of the people again, insomuch that he did blind their eyes, and lead them away to believe that the doctrine of Christ was a foolish and a vain thing." (3 Nephi 2:2). Mormon observed that "the people began to wax strong in wickedness and abominations." (3 Nephi 23). He might have been drawing particular attention to those who had made covenants with God, and who should have known better than to conspicuously compromise their standards by following Q as he meandered off the strait and narrow way into mischief. (See 1 Nephi 8:23).

It is one thing for an ignorant people to live in opposition to the laws of God, but it is quite another for those who have had the light, and who have enjoyed spiritual gifts, to turn from them, willfully rebel, and intentionally seek darkness. That scenario is abominable because it represents more devotion to Q and less faith in God. It is not easy for those who refuse to accept spiritual gifts to obtain forgiveness. Those who die without these gifts, who have not allowed Christ to work His magic in their hearts, "die in their sins, and they cannot be saved in the kingdom." (Moroni 10:26).

Anyone, and especially the Q, can easily and adroitly "teach for doctrines the commandments of men, having a form of godliness, (while denying) the power thereof." (J.S.H. 1:19). Q will always find that those with itching ears are an attentive audience. (See 2 Timothy 4:3). Someone once said that it is easy to plot a course with a map and a compass, but without a sense of who you are, you will never know if you're already home.

When Alice was in Wonderland, she asked Q, who was masquerading as the Cheshire Cat: "Would you please tell me which way I ought to go from here?" The cat responded: "That depends a good deal on where you want to go." Alice acknowledged: "I admit, I don't much care where." To which the cat retorted: "Then it doesn't matter which way you go." Alice implored: "Just so I go somewhere!" The cat coyly observed: "Oh, you are sure to do that, if you only walk far enough."

Chapter Twenty

Is God a Carbon-Based Life Form
(and other interesting trivia)

Is our Heavenly Father a carbon-based life form? We learn in the scriptures, that "there stood one among them that was like unto God, and he said unto those who were with him: We will go down, for there is space there, and we will take of these materials, and we will make an earth whereon these may dwell." (Abraham 3:24). This scripture warrants at least two pertinent observations: 1) there was "space" on the earth. This might confirm the existence of similar "space" in the eternal worlds, or it might differentiate the space found on earth from that which is found in eternal worlds, and 2) the Gods took the materials at hand, in the universe, to make an earth. A logical conclusion would be that these materials were naturally found in nature and were readily available to those charged with the creation of worlds.

Carbon must have figured prominently among those materials. It is a primary component of all life that is now found on Earth and represents up to 50% of the planet's dry biomass. It is estimated that there are 550 billion tons of carbon bound within the carbon-based life forms on Earth. Within terrestrial carbon-based life forms, its atoms form stable bonds with many elements, primarily with oxygen, hydrogen, nitrogen, phosphorus, and sulphur. Carbon is the 15th most abundant element in the Earth's crust, and the 4th most abundant element that has been found in the known universe, after hydrogen, helium, and oxygen. It is the second most abundant element in the human body after oxygen. By mass, about 96 percent of our bodies are made of the following 4 elements: oxygen (65 percent), carbon (18.5 percent), hydrogen (9.5 percent) and nitrogen (3.3 percent). These elements are found in our body's most abundant and important molecules, including water, proteins, and DNA.

In science fiction, silicon is often used as a substitute for carbon in alien life forms. In the real world, microprocessors give substance to the possibility of extra-terrestrial silicon-based life. In "Star Trek: The Motion Picture", the Ilia Probe initially tells Captain Kirk: "You are the Kirk unit. You will assist me. I've been programmed by V'Ger to observe and record normal functioning of the carbon-based units infesting U.S.S. Enterprise."

Kirk finally figures it out. "V'Ger!" he says. "VOYAGER. Launched more than 300 years ago, (in 1977), designed to collect data and transmit it back to Earth. It disappeared. Machine entities on the far side of the galaxy must have found it to be one of their own kind. They discovered its simple twentieth-century programming: To collect all data possible, and learn all that is learnable. Return that information to its creator. The machines interpreted it literally, and built this vessel so that Voyager could fulfill its programming. And on its journey back to Earth, it amassed so much knowledge, it achieved consciousness; it became a living thing."

At the end of the motion picture, the carbon-based units (Kirk, Spock, and others) realize just how the probe had adapted its mission, "to bring those who had built it here, to touch its Creator." To which, Dr. Leonard McCoy dryly observes: "To capture God! V'Ger's going to be in for one hell of a disappointment!"

The writers of Star Trek took poetic license when creating the technological intricacies of its story lines. They got many things right but took extensive liberties regarding others. But who is to say for sure whether there are other carbon-based life forms or even silicon-based sentient machine entities in the galaxy who are even now utilizing the materials at hand to invent mind-boggling systems that allow them, or their holograms, or their probes, to travel among the stars, and even to visit Earth? Among these imaginary and yet visionary technologies are the following:

<u>Androids</u> - Androids were advanced forms of A.I. that could be fitted with "positronic" brains, so that they were self-aware and could imitate nearly every behavior of humanoid life-forms. Their senses also exhibited far greater acuity than their carbon-based counterparts. Lieutenant Commander Data, posted to the Starship USS Enterprise-E after his graduation from Starfleet Academy, was undoubtedly the best-known android in the Star Trek Universe. He was created by Noonien Soong. His construction utilized tri-polymer composites, molybdenum-cobalt alloys, as well as bio-plast sheeting.

He was built with 800 quadrillion bits (800 petabytes) of storage capacity, and a computational speed of 60 trillion operations per second. These gave him abilities that far exceeded those of his carbon-based shipmates. Although he was a synthetic life form, Jean Luc Picard eulogized him after his "death" as one of their own: "In his quest to be more like us, he helped us to see what it means to be human. His wonder, his curiosity about every facet of human nature, allowed all of us to see the best in ourselves. He evolved and embraced change because he always wanted to be better." ("Star Trek Nemesis").

<u>Artificial Gravity</u> - Gravity is a property of mass. Its force is directly related to the proximity of objects that have mass, and, because of the inverse square law, with distance it drops off exponentially. In the real world, the easiest way to simulate gravity is to introduce an opposite force that is equal to its magnitude, such as by rotating a space craft as in the motion picture "2001: A Space Odyssey". The wheel needs to rotate about 9.5 times per minute (once every 6.31 seconds) to simulate 1 G. Gravity can also be simulated by linear acceleration. Theoretically, it can be produced by using incredibly dense matter for deck plating (on the order of neutron star density!)

In the Star Trek Universe, a gravity plate was used ('artificial gravity plating", "gravity net", or "synthetic gravity"). It lined the ship's hull. or was installed on interior deck plating, and was created with an artificial gravity generator that utilized duranium sheeting. The artificial gravity grid that was generated worked in concert with a graviton stabilizer that was synchronized with the warp field of the ship's engines.

<u>Cloaking devices</u> – These could render object, individuals, or even entire starships invisible. In the

real world, scientists have created crude cloaking devices, consisting of metamaterials (electronic devices that distort the reflection and refraction of light) that can hide tiny objects. Deflector shields - These relied on layers of energy distortion that were created by high concentrations of gravitons surrounding the object to be protected. The shield was emitted by a dish or grid. In the real word, research suggests that lasers might be able to mimic tractor beams, drawing microscopic objects toward the laser cavity at the source of the beam. Similar beams ("pressor beams") have demonstrated the ability to repel objects. Gravity propulsion beams have been theorized, as well as the utilization, in space, of energy from the sun that could provide a push against solar sails, thus efficiently propelling a spacecraft on "winds of light."

Dyson Sphere - A Dyson Sphere is a hypothetical megastructue that completely envelops a star to capture its solar power output. Its design attempts to explain how a spacefaring civilization could meet its energy requirements once they had exceeded the resources of its home planet.

Force Fields - Starfleet force fields were energy barriers rated by intensity (levels 1 – 10). They could be used to seal hull breaches, thereby maintaining structural integrity of the ship, and to hold intruders, assailants, or prisoners at bay. A containment field was a force field used to prevent the escape of matter or antimatter, and to isolate areas of the ship during transport. Defensive shielding was a force field that produced a bubble around a structure to be protected, especially during battle. The shielding could be weakened if it was repeatedly subjected to attack by energy weapons

Inertial Dampers - In the real world, these are devices that counter or "damp" the effects of inertia. There are many types of such devices in the real world: ballast, brakes, flywheels, counterweights, and pendulums. But in the Star Trek Universe, inertial dampers accomplished far more; they countered the effects of mind-bogglingly rapid acceleration and deceleration of starships. The inertial dampers absorbed the natural inertia of the vessel; thus, a starship could jump from a dead stop to warp speed almost instantaneously without smashing the crew (and all objects within the ship that were not bolted down) into the aft walls. Inertial dampers could also be adjusted to compensate for the gravitational forces of nearby star systems, thereby avoiding the risk of tearing the ship apart as it passed by. Matter / Antimatter – These particles are theorized to be volatile when they are in proximity to each other, resulting in mutual annihilation. Antimatter in microscopic quantities has been created in real world laboratories, but in nowhere near the quantities necessary for power generation (such as in a starship). Besides, the dilithium crystals necessary to power a warp core engine do not exist, nor do warp core engines, for that matter.

Nacelles – wings on a starship that have "Buzard scoops" to collect hydrogen atoms from interstellar space, to replenish fuel reserves while traveling at impulse (non-faster-than-light) speed.

Phasers – The acronym "Phaser" stands for PHASed Energy Rectification, but their use in Star Trek relies on fanciful physics. The only device in the real world that approaches the properties of a phaser is a stun weapon that utilizes microwave energy.

Photon Torpedoes - These were warp capable tactical matter / antimatter weapons, used in the

Star Trek Universe. Before deployment, they were contained in photon tubes. The warhead had a detonation chamber filled with antimatter. Upon detonation, a matter / antimatter explosion created a flood of ion radiation.

Silicon-based life forms – In a paper published in the peer-reviewed journal "Life", physicists from M.I.T. discussed the possibility of life on other worlds that is based on silicon rather than carbon. Although they were essentially pessimistic about the prospects, they theorized that under conditions far different from those found on earth that have been observed elsewhere in the galaxy, it might be possible that life has evolved utilizing silicon rather than carbon as its basic building block.

Tractor beams – In our future, electrically charged waves or particles (such as cosmic radiation) might be deflected by electromagnetic fields, and magnets might tow a metal object through space, but tractor beam technology such as is found in Star Trek is purely imaginary, for the time being.

Universal translators – This device made possible the instantaneous and seamless interpretation of alien languages into the native language of the listener. It utilized a "translation matrix" that eliminated the pressing need for alien races to speak or learn English. In the Star Trek Universe, virtually every alien species that was encountered seemed to already possess similar, or even conveniently identical, technology.

In addition to all the above technologies, there were many building-blocks in the Star Trek Universe that that had amazing properties but were entirely imaginary. These included:

Bio-mimetic Gel – a volatile substance with medicinal properties.

Cordrazine – a powerful stimulant used to revive unconscious patients.

Dilithium crystals – essential components of the warp drive system. They regulated the matter / antimatter reaction within the warp core itself.

Inaprovaline – a drug that restored neurological and cardiovascular functions.

Latinum – a rare liquid that was pressed with gold (a worthless element, in the Star Trek Universe) to make a stable solid that was a valuable commodity used in trade. It could not be replicated.

Omega – an unstable molecule, believed to be the most powerful substance in the universe. It could wreak havoc on subspace, making warp speed impossible.

Protomatter – a sought-after molecule, with the reputation as the best energy source in the Alpha Quadrant.

<u>Quantum Slipstream Drive</u> – a propulsion system that allowed starships to break through the warp speed barrier.

<u>Red Matter</u> – could create a black hole when it came in contact with decalthium.

<u>Rodinium</u> – the hardest substance in the galaxy.

<u>Schwartz</u> – a potent fuel that could propel a Space Winnebago incredible distances through the cosmos (in "Space Balls").

<u>Transparent Aluminum</u> – a material, whose formula Commander Montgomery Scott gave to a 20th century scientist (in order to construct a suitable tank in which whales could be transported back to the future). (In contravention to the Prime Directive?)

<u>Trans warp Drive</u> – utilized a new form of dilithium, discovered by Voyager when it was in the Delta Quadrant, that retained stability at speeds exceeding warp 9.7. Lt. Tom Paris designed a depolarization matrix around the fuselage, to take the ship to warp 10.

<u>Trellium-D</u> – an alloy used to protect starships against spatial anomalies.

<u>Trilithium</u> – a compound capable of stopping fusion within a star. It inhibited nuclear reactions.

<u>Tritanium (and Duranium)</u> – the principal building blocks of space structures, such as bulkheads, with a hardness off the standard 1 – 10 scale.

<u>Unobtanium</u> – In the motion picture Avatar, unobtanium was a valuable mineral found on the moon Pandora.

<u>Verterium Cortenide</u> – the only substance known to be capable of generating a warp field, when supplied with energy, in the form of plasma from the warp core. Thus, warp coils were constructed from this material.

These technologies and materials from the Star Trek Universe and elsewhere in science fiction are illustrations of our imaginations, but they stretch our minds and open the door to possibilities that reside somewhere beyond the borders of theoretical physics. Who can say whether interstellar Type 3, 4, or 5, or Omega Minus alien civilizations are already familiar with them, and utilize their unique properties to wander among the stars? If so, the basic rules that we have established regarding interstellar travel and communication between species might not apply. First Contact might be closer than we think.

Chapter Twenty-One

Is There a Prime Directive?

In the fictional universe of Star Trek, the Prime Directive, Starfleet's General Order #1, is the most prominent guiding principle of the United Federation of Planets. It dictates that there can be no interference with the internal development of pre-warp civilizations, consistent with the historical real-world concept of Westphalian sovereignty.

The Prime Directive emphasizes that civilizations with advanced technologies should not adjust, alter, amend, change, develop, expand, improve, modify, or in any way revise the natural development of emerging pre-warp societies, wherever they may be found, even if the aid is well-intentioned or kept secret. To do so might radically influence the natural evolution of such a civilization. This could be detrimental to the society of those celestial neighbors within the sphere of the Federation's expanding power, or it could beneficial, but the effect would most certainly not be neutral. In any case, the culture's natural progression, and that of those with whom it might later come in contact, would be artificially influenced in ways that would be difficult to anticipate and impossible to control. No matter the outcome, there would be no going back.

The Prime Directive works well for starships boldly going where no one has gone before, but it runs counter to the nurturing philosophy of our Heavenly Father. His stated mission is to care for His children to the extent that they might flourish under His active guidance until they reach the point where they will receive everything He is and has. For all practical purposes, and in Star Trek terms, they will have achieved warp capability, and will be able to travel among the stars, under His continuing guidance, at faster than light speed. (See Chapter Seven: Beings of Light). No matter in what condition they may currently find themselves, at some point, He will provide the means for them to be reborn by generation and not simply by maturation, to receive immortality and eternal life. His version of The Prime Directive has been radically altered and begins with: "Come follow me" because He is our Exemplar. (Luke 18:22, see 1 Peter 2:21).

In the real word, we can only indirectly appreciate that Heavenly Father is the President of a loosely defined and somewhat disparate United Federation of Planets. However, His agenda is radically different from that of Starfleet. He enjoys life tenure, and has at His disposal skills, knowledge, expertise, know-how, tools, and technologies that are beyond our comprehension. Were they to be distributed among us all at once, however, they would probably do more harm than good. It would be of no benefit for us to receive a cascading cornucopia of information or technologies that would simply overload our processing capacity. So, instead of overwhelming us, and at the same time in direct contravention not only to the tenets of the Prime Directive, but also to the Temporal Prime Directive, He gives us "line upon line, precept upon precept; here a little, and there a little; giving us consolation by holding forth that which is to come, confirming our hope!" (D&C 128:21).

Sir Isaac Newton recognized his woeful inadequacies, writing late in his life: "I do not know what I may appear to the world, but to myself I seem to have been only like a boy playing on the sea-shore, and diverting myself now and then in finding a smoother pebble, or a prettier shell than ordinary, while the great ocean of truth lay all undiscovered before me." ("The Ascent of Man," p. 236). Starfleet's Prime Directive would forbid our exposure to that vast ocean that lay before Newton's eyes, but God's greater vision invites us to prepare ourselves, as Q told Captain Picard, to partake of wonders more incredible that we can possibly imagine, and treasures to satiate desires both subtle and gross. If er also encounter terrors to freeze our souls, at least we have the consolation of hope. (See 2 Thessalonians 2:16). Life is not for the timid!

God understands that our minds are generally locked on telestial targets, and that, when we attempt so-called higher-level thinking, we risk becoming "as sounding brass, or a tinkling cymbal." (1 Corinthians 13:1). "For my thoughts are not your thoughts," the Lord revealed, "neither are your ways my ways ... For as the heavens are higher than the earth, so are my ways higher than your ways, and my thoughts than your thoughts." (Isaiah 55:8-9). Because that is so, to ensure the expansion of our understanding, He made it possible fore us to be carried along on currents of insight, intuition, inspiration, discernment, and revelation, buoyed by the hope that "the best of life is always further on. Its real lure hidden from our eyes somewhere behind the hills of time." (Sir William Murdock). It is our faith that good things come to those who are willing to wait, and it is our testimony that Father knows best. (See James 5:7 & Moses 5:11).

Our divine tutorial training is on-going process. As we receive guidance, we "hear truth spoken with clarity and freshness; uncolored and untranslated, it speaks from within ourselves in a language original but inarticulate, heard only with the soul." (Hugh B. Brown). Our Father made sure that the Light of Christ and the Holy Ghost would always be available to educate us about eternal truths. They continue to teach us things that we can learn in no other way. As Tom Bodett famously said, so could it be said of the Light of Christ and the Holy Ghost: "We'll leave the light on for you."

"Hearken unto me" urged King Benjamin, as you are taught the will of the Lord, the mind of the Lord, the word of the Lord, the voice of the Lord, and the power of God unto salvation, "and open your ears that ye may hear, and your hearts that ye may understand; and your minds that the mysteries of God may be unfolded to your view." (Mosiah 2:9, see D&C 68:4). When we allow the Light of Christ, and especially the Holy Ghost, to influence our lives in quiet whisperings or in words spoken by those who are under their influence, we are cast off into streams of revelation and carried along in the quickening currents of direct experience with God. The philosophy of The Prime Directive is annulled by a flowchart whose tenets are founded upon "the unspeakable gift of the Holy Ghost" and whose principles are fundamental to greater understanding that comes via the participatory Plan of Salvation. (D&C 121:26).

That involvement is just how God wants it to be. In effect, He says, "Stop thinking about The Prime Directive." Forget the crooner who sings: "And now, the end is near, and so I face the final curtain. My friend, I'll say it clear; I'll state my case, of which I'm certain. I've lived a life that's full. I've traveled each and every highway; and more, much more than this, I did it my way." ("My Way," lyrics by Paul Anka). What a pathetic waste, to be so arrogant, haughty, egotistical, and proud to summarily decline and publicly repudiate the helping hand of God whenever, wherever, and however it is offered.

"Precious Lord, take my hand, lead me on, let me stand. I am tired, I am weak, I am worn. Through the storm, through the night, lead me on to the light. When my way grows dreary, precious Lord, linger near me. When my life is almost gone, hear my cry, hear my call, hold my hand lest I fall. When the darkness appears and the night draws near, and the day is past and gone, and at the river, I stand. Guide my feet and hold my hand." (Thomas Dorsey).

When we submit ourselves to instruction by the Spirit, we are reminded that "we can make our lives sublime, and, departing, leave behind us footprints on the sands of time." (Henry Wadsworth Longfellow, "A Psalm of Life"). Sometimes, there are two sets of footprints that trace their way along the shoreline of our personal discovery, and during trying times, there may be only one set. Of course, it is during those challenging times that the Savior has carried us on His shoulders. We are comforted that He will lead us away from stormy seas, and out of the bondage of ignorance "by power, and with a stretched-out arm." (D&C 103:17). The very fact that He is "mighty to save" gives us confidence that, ignoring the Prime Directive, He will actively intervene in our lives, and rescue us from our own worst enemies. (2 Nephi 31:19).

For the many millions of people who don't consciously recognize their need for God's help, the faith of their fathers may still consist "of a humble admiration of the illimitable superior spirit who reveals himself in the slight details we are able to perceive with our frail and feeble minds." (Albert Einstein). The great mathematician also said: "I am satisfied with a glimpse of the marvelous structure of the existing world, together with the striving to comprehend a portion of the Reason" or, in other words, the Intelligent Design, "that manifests itself in nature." ("The World as I See It"). The momentum of God's Divine Design carries us beyond the Prime Directive by nurturing our relationship with the power of the Infinite.

Einstein's unknowable Reason notwithstanding, it is the heartstrings of artists that more frequently vibrate in harmony with the rustling robes of the Spirit. As the poet Eliza R. Snow wrote: "Oft times a secret something whispered, 'You're a stranger here,' and I felt that I had wandered from a more exalted sphere." (O My Father). She had experienced God's intervention in her life, Starfleet's General Order #1 notwithstanding.

So, too, had Ralph Waldo Emerson, whose testimony was: "Those who have seen the rising moon break out of the clouds at midnight have been present like an archangel at the creation of light and of the world." By the power of the Light of Christ, he had been taught about the relationship between light and truth, and the Holy Spirit. When God said, "Let there be light," it was a simple statement of fact as much as it was a command. It was an invitation to come in out of the cold to bask in His glory, to embrace truth, and to celebrate light; to be energized by it, to be washed clean in its influence, and to revel in it.

As are most of His entreaties, the formula is simple. Because He has manifested Himself to us in so many easily recognizable ways, He knows how we will respond, when He asks: "What think ye of Christ? Whose son is He?" (Matthew 22:42, see Chapter Two: Love Letters from God). He knows that because we have pondered the solemnities of eternity we will see evidence of His hand everywhere, and especially in nature; that "the earth is crammed with heaven, and every bush with fire of God. But only those who see, will take off their shoes. The rest will stand around picking blackberries." (Elizabeth Barrett Browning, see D&C 43:34).

We are intertwined with God in a palpable connection, for we are His predilected people, and we live within his embrace, enjoying security that others do not know. He takes notice of each of us, just as He does of sparrows falling from trees, and supernovas exploding in distant star systems. We can rest assured that, when we make First Contact with alien species, it will be obvious that God has not played dice with His creations. First Contact will confirm that His nations are numberless, as are the inhabitants of distant worlds. It will validate the principle that He remembers those who are upon "the isles of the sea." First Contact will expand our borders of our comprehension to include all those throughout the vast ocean of the cosmos, and upon all beaches of all the earths that lie beneath the heavens. (See 2 Nephi 29:7, and Chapter Eleven).

On none of these creations does He leave things to chance. As a loving Parent, He actively nurtures us, no matter where we may be scattered across the galaxy. He always attends our Parent/Teacher conferences. He never misses our extracurricular activities. He has a season pass to every game we play. He is forever in the bleachers, right in the middle of the cheering section. He is in the "wave" when it passes through the stands. He sits up late in the evening, with the porch light on, waiting for us to come back home to safety. He has our phone numbers on speed dial, and regularly uses voice mail, text messaging and Find My Friends. We are at the top of His Favorites list.

But He does not participate in Hide and Seek or Capture the Flag. Even if we try to ignore Him, we cannot make Him go away. We can hear Him now: "This is My Prime Directive", Celestial Kingdom General Order #1, to bring to pass the immortality and eternal life" of My children." (See Moses 1:39). He will never leave us alone. He "will go before (our) face. (and) will be on (our) right hand and on (our) left, and (His) Spirit shall be in (our) hearts, and (His) angels round about (us), to bear (us) up." (D&C 84:88). He" will never leave (us), nor forsake" us. (See Hebrews 13:5).

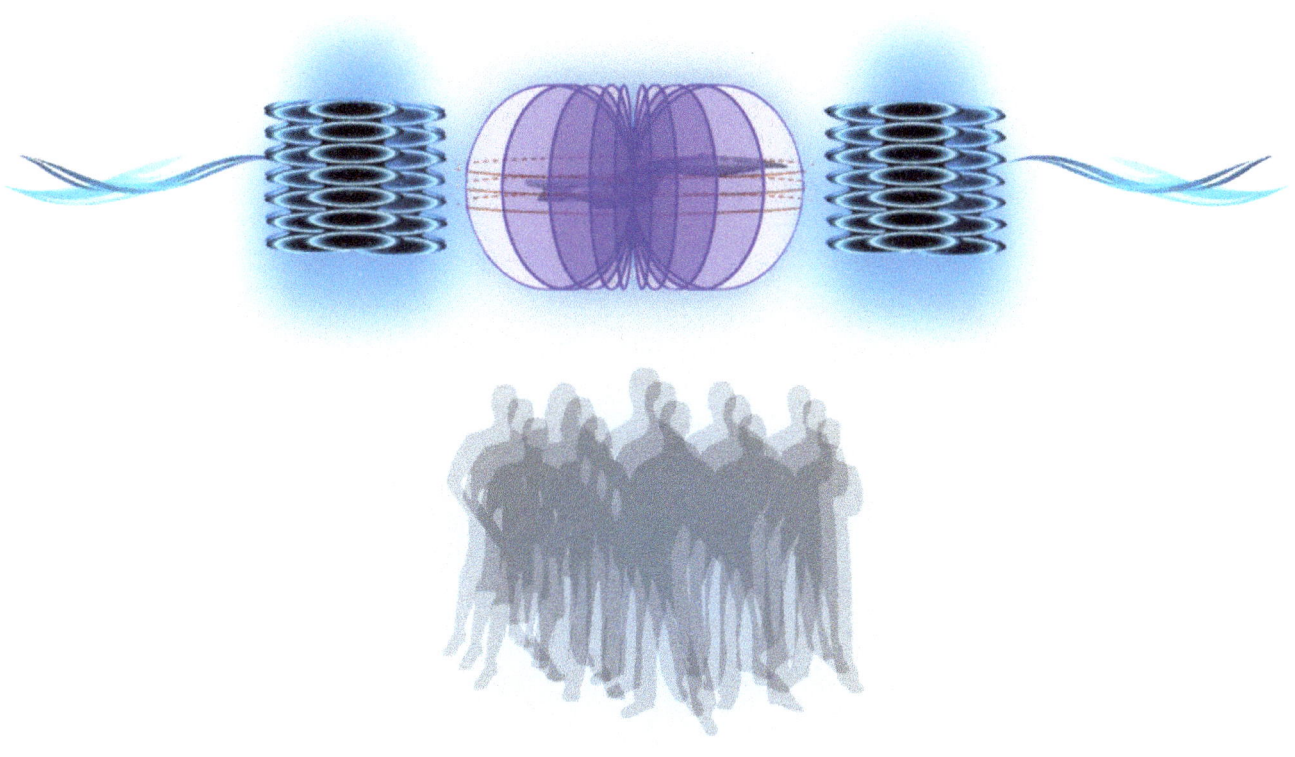

Chapter Twenty-Two

Our Genetic Code Has Been Scattered Across the Universe

The story is told of young Spencer W. Kimball traveling alone by train from his home in Thatcher, Arizona, to Salt Lake City. An older man climbed aboard at an intermediate stop and sat down next to the 10-year-old boy. He asked: "Where are you headed?" "Salt Lake City," was his reply. The next question was predictable: "Are you a Mormon?" "Yes," answered Spencer.

Then the man asked the obvious question: "What do Mormons believe?" Spencer had been grounded in the Gospel by faithful parents who had taught him the thirteen Articles of Faith, and so he answered: "Well, we believe in God the Father, and in His Son Jesus Christ, and in the Holy Ghost." "What else do you believe?" asked the man. "We believe that we will be punished for our own sins, and not for Adam's transgressions." And so it went.

I was reminded of President Kimball's experience as I sat on the stand at the funeral of a faithful ward member. I had been asked to give the eulogy, and as I looked out over the congregation, I saw unfamiliar faces, many of whom were the non-member relatives of my friend. Then, as I stood before them, I simply said: "Claude was a good man, who believed in our Father in Heaven. He had a firm an abiding testimony of the divinity of His Son Jesus Christ, and he regularly felt the influence of the Holy Ghost. He believed that we will be punished for our own sins, and not for the transgression of Adam, and thus, he was grateful for the Atonement of Jesus Christ." And so it went.

As our children were growing up, their mother and I always told them that if they memorized the Articles of Faith they would never be without the material for an extemporaneous address in Church. We made it a point in our family to make sure that they knew them by heart before their baptismal interview with our Bishop. We taught them to make the articles of their faith the tangible particles of their faith. In doing so, we almost unconsciously helped them construct the foundation teaching upon which they would build their lives: that they were children of our Heavenly Father, Who knew them, loved them, and watched over them. Their memorization of the Articles of Faith helped to energize within their DNA a genetic code that has been scattered across the cosmos, that we are children of our Heavenly Father.

Consider the scenario at a typical social gathering, where we meet a total stranger. Within 2 minutes, we are on a first-name basis, and our new friend is soon asking: "What do you do?" What they mean is: "What do you do for a living?" That is interesting, since we do many things. We identify ourselves in myriad ways, including the place of our birth, our nationality, the location of our residence, our dialect, our interests and hobbies, our alma mater, our favorite sports teams, our political, fraternal, and religious affiliations, and yes, even our occupation. None of these descriptions are wrong, unless they supersede questions relating to the genetic code that binds us to the cosmos, and to our Heavenly Father.

I am not suggesting that we should lead with: "Hello. I am a child of God." But we might want to steer away from rote answers that are superficial, have little meaning, create negligible impact, and leave our listeners with scant information of real value. If our "First Contact" with our new friends is framed around who we are, and who they are, that we are all sons and daughters of God, we may be better prepared for the Spirit to then work its magic. The Lord has promised: "I will go before your face. I will be on your right hand and on your left, and my Spirit shall be in your hearts, and mine angels round about you, to bear you up." (D&C 84:88). This close association with our Heavenly Father is a principle that maintains its validity as we establish First Contact with His children at social gatherings in our neighborhood, and it will serve us well if we are called upon to introduce ourselves at interstellar rendezvous points here on earth, or at distant assemblies across the far reaches of the galaxy.

The lyrics to the song "I am a Child of God" were written in 1957 by Naomi Randall (1908 - 2001). A friend composed the music. It is one of 45 hymns that The Church of Jesus Christ of Latter-day Saints includes in its basic musical curriculum, and it is one of the first hymns that new members learn. It has been translated into over 90 languages, and its lexicon has become a means of teaching a doctrine that is simple and easy to understand. Thus, it is frequently found in church instructional curricula, it is woven into sermons, and it even finds itself the subject of merchandizing and novelties. It is a song that is easy to understand, because it transcends time and space. Its resonant theme has stirred the hearts of countless of God's children, no matter where they may be busily engaged in their mortal curriculum. (For some of our interstellar cousins, it would help to have a Universal Translator handy).

Randall composed the song at the request of the Primary General Board, of which she was a member at the time. Its objective was to have a tune that Primary age children could easily remember and sing, that would reflect doctrinal teaching on the nature of our relationship with our Heavenly Father. Randall described how she went about fulfilling her commission from the Board: "I got down on my knees and prayed aloud, pleading that God would let me know the right words. Around 2:00 a.m., I awakened and began to think again about the song. Words came to my mind, and I immediately got up and began to write them down as they had come to me. Three verses and a chorus were soon formed. I gratefully surveyed the work, drank of the message of the words, and returned to my bedroom, where I knelt before my Father in Heaven to say: Thank you!" In the morning, she mailed the lyrics to her friend Mildred T. Pettit, who immediately wrote the accompanying music.

Sister Randall had been guided by the Spirit, as had Job of old. He described how "in a dream, in a vision of the night, when deep sleep falleth upon (us, and) in slumberings upon the bed. Then he openeth (our) ears, and sealeth (our) instruction." (Job 33:1-16).

"I am a Child of God" was first performed at a stake Primary conference in 1957. After hearing it, that same young boy from Thatcher, Spencer W. Kimball, who by then was an Apostle, asked the Primary General Board if the phrase "Teach me all that I must know" could be changed to "Teach me all that I must do." As he later explained, "To know isn't enough. The devils know and tremble. We have to do something." The suggestion was gratefully accepted by Sister Randall, and the change was made.

"I am a Child of God" was first published in the "Sing with Me" songbook for children. (1969). In 1978, Sister Randall composed a fourth verse, but when the song was added to the L.D.S. Hymnal that same year, that verse was excluded because it had not been an official part of the original song. However, in 1989, when a new songbook for children was published, the fourth verse was included.

Today, "I am a Child of God" (with three verses) is selection number 301 in "Hymns of The Church of Jesus Christ of Latter-day Saints," and the song is included in the "Children's Songbook" with the 4th verse included. This brings us to the first verse: "I am a child of God, and He has sent me here; has given me an earthly home, with parents kind and dear." As William Wordsworth wrote: "Our birth is but a sleep and a forgetting. The soul that rises with us, our life's star, hath had elsewhere its setting, and cometh from afar. Not in entire forgetfulness, and not in utter nakedness, but trailing clouds of glory do we come, from God, Who is our Home." (Ode: Intimations of Immortality, from Recollections of Early Childhood).

Our understanding that God's genetic code has been scattered across the galaxy places a high priority on our responsibility here on Earth to teach children correct principles, thereby to nurture their faith in the titular Head of our extended family. "Faith," after all, "cometh by hearing, and hearing by the word of God." (Romans 10:17). "For unto us," wrote Paul, "was the gospel preached, as well as unto them (of ancient Israel); but the word preached did not profit them, not being mixed with faith in them that heard it." (Hebrews 4:3).

In the Book of Moses, we learn that the great lawgiver's sense of identity was based on his eternal relationship with his Father in Heaven. (Moses 1:3-4). He knew that he was God's offspring. In fact, he was inexorably intertwined, in a double helix, with God. That bond was fixed and immovable, and it could not be broken, and when his Father spoke to him, He confirmed: "Behold, I am the Lord God Almighty, and ... thou art my son. (Moses 1:3-4). Today, we teach our children who they are, and the missionaries cite these same scriptures that resonate with truth. That burgeoning knowledge largely defines our character, and the strength of our conviction firmly establishes our place in the cosmos, shoulder to shoulder, as it were, with all our brothers and sisters who reside upon God's creations.

And then, the chorus: "Lead me, guide me, walk beside me, help me find the way. Teach me all that I must do to live with Him some day." Children learn to be obedient, that one day they may, in turn, teach their own offspring, in an unbroken pattern. But if there are children in Zion whose parents fail to teach them to understand the doctrines of the kingdom. and especially the truth that each of us is a star-child, trailing clouds of glory, as it were, from God, Who is our Home, "the sin be upon the heads of the parents." (D&C 68:25-28).

The second verse reads: "I am a child of God, and so my needs are great. Help me to understand his words before it grows too late." The three most important days of our lives are the day we were born, the day we find out why, and the day we die. With that awakening awareness, we come full circle. "When a baby is born, and as we wait with those who are dying, we brush against the veil, as greetings and goodbyes are said almost within earshot of each other. In such moments, this resonance with realities on the other side of the veil is so obvious that it can be explained in only one way." (Neal A. Maxwell, "B.Y.U. Devotional," 11/1979).

All the children of God will make a giant leap in time, space, and faith, when they return to Him one day. We can almost hear the angelic voices now: "Here you are, home from your mission. It seems like it was such a short time. Think of the people you met, the people you helped. Think of how you have grown spiritually. You were just a child when you left home, not so long ago. There is Mother waiting to embrace you, standing just a bit behind Father, who is bursting with pride. Are those tears of happiness on Mother's cheeks? Father first strikes hands with you, and then pulls you into His embrace. The feelings are resonant, and you know this is where you belong – this is a real homecoming – home to Heavenly Father and Mother." (Anonymous).

The third verse reads: "I am a child of God. Rich blessings are in store. If I but learn to do his will, I'll live with him once more." Alma called God's Plan "The Plan of Happiness." (Alma 42:16). Its design provides a way for us to acknowledge the genetic blueprint of our lives and follow it past solar systems, through star clusters, around nebulae, and beyond galaxies, to find eternal happiness,. God's Rest can be pinpointed at the very center of our star charts, and "is the object and design of our existence and will be the end thereof if we follow the path that leads to it. And this path includes faith, virtue, uprightness, and keeping all the commandments." (Joseph Smith, "Teachings," p. 255-256).

As we engage the curriculum of life's learning laboratory, it helps to have celestial signposts and millennial mile markers to navigate the telestial traffic jams, the conceptual cul-de-sacs, and the doctrinal deviations that threaten to detour us from the strait and narrow way onto telestial turf and the minefields of mortality. The expanding circle of opportunity afforded by obedience to Gospel principles, however, assures us of direct experience with the perfect law of liberty. Thus, we trade the uncertain course adopted by individuals bound for the telestial kingdom for the certain reality of celestial surety. That knowledge is as a warning buoy, an aid to navigation for the children of God, no matter in what part of the cosmos they may reside.

In countless episodes of "Star Trek: The Next Generation", Captain Jean Luc Picard sat in his chair on the bridge of the Enterprise-E, raised his right hand, pointed two fingers toward the view screen, and uttered the command: "Engage!" With that, off they went in their starship, inertial dampers online as they almost instantly reached maximum warp. We do something similar when we engage our agency, which is to us as our own personal Quantum Drive propulsion system. Our course is laid in, and the parameters of our journey are defined by our Operations Manual, whose elements can be found within The Plan of Salvation. The only boundaries it recognizes are those the Lord has set. Our navigational deflector shield is Jesus Christ, and our inertial damper is His Atonement. Our willingness to "engage" is determined at the molecular level by our genetic code. Its blueprint will take us to the final frontier of experience at the edge of eternity, where we will encounter strange new worlds, new life, and new civilizations, in regions of space where no one has gone before.

The good decisions we make along the way, during our visionary travels across the galaxy, will take us to intermediate "star bases", and particularly to the Sacrament table. These stops will automatically negate the consequences of poor choices that sometimes creep into our travel itinerary. Our inertial damper will protect us from the spatial distortions of sin, from the negative effects of inertial indecision, and from the buffetings of the gravimetric waves of unlimited freedom that would lead us to the temporal distortions of unrelenting tyranny. On our journey to the far reaches of our imaginations, even without the benefit of a trans-warp conduit, we will be reassured by the knowledge that we are children of God, proven by a heraldry that proclaims our independence in that stage of development to which our decisions have led us. Our divine genome will become an armorial bearing that distinguishes us as members of a royal family. In this, the best of circumstances that are dictated by the principles of The Plan, "the universe itself will become a machine for the making of Gods." (Henri Bergson, "Two Sources of Morality and Religion," p. 306).

There is a fourth authorized verse of "I Am a Child of God" that is included in the Children's Songbook: "I am a child of God. His promises are sure. Celestial glory shall be mine, if I can but endure." To know that we are children of God is enough to kindle a spark within us that ignites our sense of wonder, illuminating our understanding of the depths of eternity. It gets the dilithium crystals in the warp drive of our resolve up to critical mass. But it would be wrong to leave it at that, for to know only would underestimate the magnitude of The Plan and put at risk our relationship with God in such a way that its overarching importance in every aspect of our lives might be diluted or demeaned. The matter / antimatter interaction in our warp core might overload, and our genetic code might be misinterpreted, underutilized, or even ignored.

At the end of the day, we need to know that we are God's children, and then act upon that innervating knowledge. Doing so frees us to keep our finger on our quickening pulse, as we actively monitor our burgeoning relationship with our Heavenly Father. Our faith is founded upon the very points of doctrine that address salvation and exaltation, and upon these elements hinges its correct understanding. "I am a Child of God," as it turns out, is a very good choice of words when describing the intimacy that our Heavenly Father desires to have with us, and with His children everywhere.

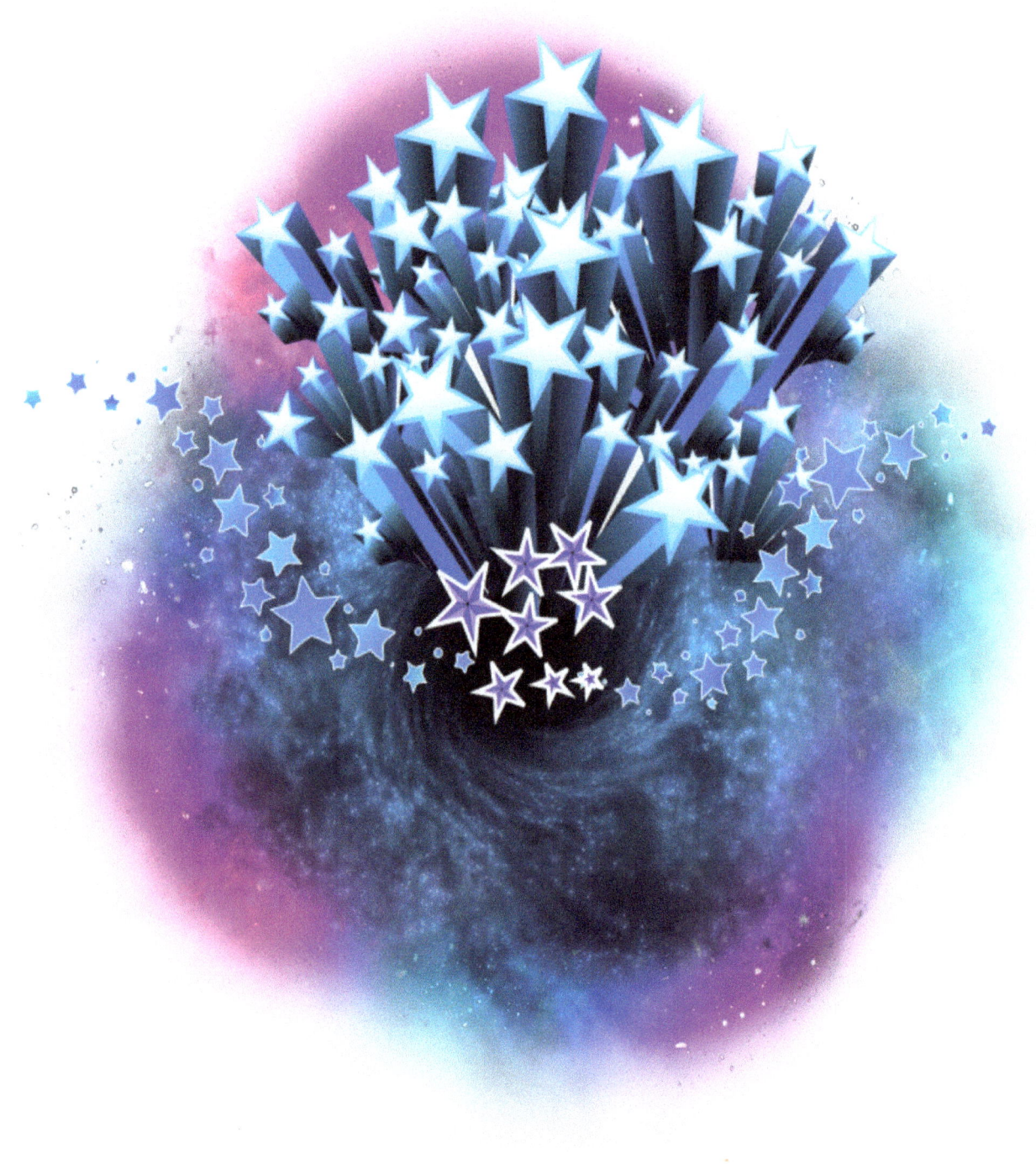

Chapter Twenty-Three

The Unknown Possibilities of Existence

We left our heavenly home to chart the unknown possibilities of existence, and we have come to Earth "like gentle rain through darkened skies, with glory trailing from our feet, and endless promise in our eyes. We are strangers from a realm of light, who have forgotten all - the memory of our former life and the purpose of our call. And so, we must learn why we're here, and who we really are." (See "Saturday's Warrior," lyrics by Doug Stewart).

Life's greatest questions plumb the depths of the unknown possibilities of existence. Where did we come from? Why are we here? Where are we going? There are about 3,300 questions in the Bible, and many relate to these three, including "Adam, where art thou?" (Genesis 3:9). "Where is he that is born king of the Jews?" (Matthew 3:2). "Which is the greatest commandment in the law?" (Matthew 22:36). "Who is my neighbor?" (Luke 10:29). "Am I my brother's keeper?" (Genesis 4:9). "If a man die, shall he live again? (Job 14:14). "If God is with us, who can be against us?" (Romans 8:31). "What must I do to be saved?" (Acts 16:30). "What think ye of Christ? Whose son is he?" (Matthew 22:42).

Substantive answers require that we embark upon personal journeys that reflect the on-going mission of the Starship Enterprise: "To explore strange new worlds, to seek out new life and new civilizations, to boldly go where no-one has gone before," and even to chart the unknown possibilities of existence. We are reminded of Dag Hammarskjöld's observation that "the longest journey is the journey inward, for he who has chosen his destiny has started upon a quest for the (divine) source of his being."

When we chart the unknown possibilities of existence, we sweep aside the self-limiting belief that "the sky is the limit," and we substitute the mind and soul expanding certainty that "heaven is the limit." This is in keeping with one of the greatest contributions of Joseph Smith, namely, his "knowledge of what is to come after death. He did much to clarify our understanding of heaven, and to make it seem worth working for." ("My Religion & Me" Course Manual).

When we open our minds to unknown possibilities, to options we have never considered, we envision a special place called Kolob, "signifying the first creation, nearest to the celestial, or the residence of God." (Facsimile #2). Of our relationship to that realm, William W. Phelps wrote: "No man has found pure space, nor seen the outside curtains, where nothing has a place." In the star-studded matrix within which he imagined Kolob, there was no end to matter, space, spirit, or race; virtue, might, wisdom, or light; union, youth, priesthood, or truth; glory, love, or being; because these are markers on the star chart upon which are revealed the bounds and conditions of unknown worlds. (See: William W. Phelps, "If You Could Hie to Kolob").

A point of reference like Kolob grounds us to certainties, even if they remain elusive. As we chart the unknown, we struggle to wrap our minds around an expansion of knowledge that doubles every 12 months, and with the realization that there is no way for us to keep up. If we do not stay focused on Kolob, we risk succumbing to the pessimistic observation that not only has knowledge outpaced truth, but that truth is having a hard time even holding its own. Knowing that Kolob exists gives us a measure of hope that we will be able to distinguish between knowledge and wisdom, and to make correct choices based on the intelligent application of the former, that we might abundantly experience the latter.

Accepting the challenge to expand our minds and our horizons, and to probe the depths of the unknown, forces us to ask ourselves difficult questions: Have we embraced the moral element of responsibility that goes hand in hand with knowledge? Do we have the spiritual and intellectual fortitude to temper knowledge with accountability in the cold light of day? When we dare to grapple with these interrogatives, we come to an epiphany, as we determine to do our best to be righteous stewards. It was with this in mind that Joshua challenged Israel: "Choose you this day whom ye will serve." (Joshua 24:15).

We realize that we have the privilege to be bathed in the matter / antimatter warp core of an innervating vitality, and to be empowered with the omega particle of an otherworldly serenity. As Bagheera, the powerfully built black panther confided to Mowgli the man-cub: "I had never seen the jungle. They fed me behind bars from an iron pan until one night I felt that I was Bagheera the Panther, and no man's plaything, and I broke the lock with one blow of my paw, and I came away." (Rudyard Kipling).

We are voyagers who have already embarked upon a vast ocean to courageously sail to the edge of the world. We have embraced the task that lies ahead, because we know that our undertaking is consistent with our Heavenly Father's mission statement to bring to pass our immortality and eternal life. (See Moses 1:39). We are enlightened explorers, and the Holy Spirit has given us fire for the deed. Each of us has determined to use the 3.3 pounds of cortical grey matter with which we have been endowed (consisting of around 100 billion cells with 100 trillion neural connections) to good advantage. Such a breathtaking network blesses us with enough resources and to spare, to expand our minds and our horizons. God has clearly provided sufficient wiggle room between our ears to allow us to do so within the arena of free will. He has created the means for us to confidently step out off the edge of forever into swiftly flowing currents that will carry us onto the unexplored ocean of eternity, that we might one day be cast upon a far shore to discover the unknown possibilities of existence. "There is a tide (after all) in the affairs of men which, taken at the flood, leads on to fortune. Omitted, all the voyage of (our lives) is bound in shallows and in miseries. On such a full sea are we now afloat, and we must take the current when it serves, or lose our ventures." (Shakespeare, "Julius Caesar").

In the process, we must abandon the idolatry that obstructs our vision, and we must conquer our self-deification. We must forsake the worship of our own creations, and liberate ourselves from the cult of the state, from avarice, and from our lust for power and domination Our mission statement will transcend that of Starfleet, which, although noble in purpose, fails to recognize its moral obligations as transcendent and divine. During our journey to chart the unknown possibilities of existence, we will defend our homes from aggression and do our best to maintain peace and freedom within our borders, as does the Federation. But we will bring to our efforts a heavenly perspective as we explore strange new worlds, seek out new life, and new civilizations, and boldly go where no one has gone before.

Along the way, our physical well-being will not save us, because what is at stake is feeling and not knowledge. As we continue our long journey of discovery, our hearts and our nature will change, the scales will fall from our eyes, and the path before us will be brightly illuminated, so that we may see with the eye of faith all the way into eternity. God stands ready to bestow upon us that perspective, and only waits upon our initiative before He acts in our behalf, to make smooth the rough places in our path. (See 1 Nephi 17:46).

In the Star Trek episode, "All Good Things", Captain Jean Luc Picard asks Q: "What is it you're trying to tell me? To which, Q tantalizingly replies: "You'll find out."

Chapter Twenty-Four

Let There Be Light

"In the beginning, God created the heaven and the earth. And the earth was without form, and void; and darkness was upon the face of the deep. And the Spirit of God moved upon the face of the waters. And God said, Let there be light; and there was light. And God saw the light, that it was good: and God divided the light from the darkness. And God called the light Day, and the darkness he called Night. And the evening and the morning were the first day." (Genesis 1:1-5).

Light is a fascinating physical property. Natural light is necessary for brain development and maintenance, and no amount of artificial light can change this. Natural light is needed for mood stabilization, a healthy sleep-wake cycle, and for boosting the immune system through vitamin D production.

It is interesting to speculate about how we, or extra-terrestrials, might deal with the need for natural sunlight on an extended voyage through interstellar space. Living on a starship with no natural light for several months or years would almost certainly make life difficult. A holodeck, imitation sunlight, or shore leave on a M-Class planet orbiting a star similar to our Sun might help mitigate the problem. But, as most of us realize, there's nothing quite the same as lying on a tropical beach, soaking up the sunshine.

The Light of holiness is an entirely different matter, and it is infinitely more important when attempting to maintain balance in our lives. It is the power that permeates the universe, and Jesus Christ is the Architect of the cosmos, including the "Pillars of Creation," elephant trunks of interstellar gas and dust in the Eagle Nebula, 7,000 light years from Earth. In an 1857 sermon entitled "The Condescension of Christ," London pastor Charles Spurgeon coined the phrase to describe both the physical world and the force stemming from the Divine that binds it all together. "Now wonder, ye angels," Spurgeon wrote of the birth of Christ, "the Infinite has become an infant; He, upon whose shoulders the universe doth hang, nurses at his mother's breast; He who created all things and bears up the Pillars of Creation."

Wrote the philosopher: "There is not enough darkness in all the world to put out the light of even one small candle." The Savior taught: "The light of the body is the eye. If, therefore, thine eye be single, thy whole body shall be full of light ... and there shall be no darkness in you, and that body which is filled with light comprehendeth all things." (3 Nephi 13:22 & D&C 88:67).

The words of Christ must have penetrated the Nephites to the core, as they groped about within the darkness that enveloped the land after the crucifixion. It was so profound that not even a fire could be kindled. (See 3 Nephi 8:21). From the heavens above, He declared: "I am the light and the life of the world." (3 Nephi 9:18). Eighteen hundred years later, He told Joseph Smith that he, too, must "be a light to the world." (D&C 45:9). The light of Jesus Christ that is reflected on the countenances of the righteous is a beacon that unerringly guides those who are seeking the truth, upon whatever shores they might have been cast within His endless creations. That light guides them to the safe harbor of the Gospel, even from the farthest reaches of a vast galactic ocean. In contrast, those who are in the grasp of darkness, who refuse the Gospel, may as well be adrift in the cold vacuum of space. Their habitations will eventually become desolate, forlorn, and forsaken, as nature withholds her bounties. If we alienate ourselves from God, all the world becomes our enemy, and the Milky Way itself becomes a dark and foreboding expanse, a raging interstellar sea and a galactic graveyard with fearful solar flares, terrifying supernova, deadly bursts of gamma rays, strange magnetars, rogue black holes, and cosmic cannibalism.

In the days of Enoch, when he spoke the word of the Lord by the power of the priesthood, "the earth trembled, and the mountains fled, even according to his command; and the rivers of water were turned out of their course; and the roar of the lions was heard out of the wilderness." (Moses 7:13). The commotion in the world that Moses described was the terrestrial manifestation of cosmic confusion amid galactic turmoil.

Joseph Fielding Smith cautioned the Saints: "We should wake up to the realization that it is because of the breaking of covenants, especially the new and everlasting covenant, which is the fulness of the Gospel as it has been revealed, that the world is to be consumed by fire and few men left. Since this punishment is to come at the time of the cleansing of the Earth when Christ comes again, should not Latter-day Saints take heed unto themselves? We have been given the new and everlasting covenant, and many among us have broken it, and many are now breaking it; therefore, all who are guilty of this offense will aid in bringing to pass the destruction in which they will find themselves swept from the Earth when the great and dreadful day of the Lord shall come." ("Deseret News," 10/17/1936).

Dark matter may account for 85% of the material substance of the known universe, but the light and life of the cosmos is Jesus Christ. He is everlasting, and His influence extends from beyond a land before time, through our developmental nurturing as spiritual children of our Heavenly Father, on into mortality with its twists and turns, sunshine and shadows, and light and darkness, and finally to our glorious reunion with Him in the dazzling celestial luminosity of the eternities. From our perspective, His light has always existed. His flawless character has always focused and clarified light as does a magnifying lens. In the best of circumstances, we learn to reflect the light of His faultless and unblemished character. Because He illuminated a world that had been enshrouded in spiritual darkness, Mormon was prompted to declare: "In Christ there should come every good thing." (Moroni 7:22).

His light is a Type of completeness, bathing our mortal experience in a dazzling physical and spiritual aether that becomes an animated culture medium that fosters progress. Through a bond that is made possible by the Holy Ghost, it washes over us as a metaphysical agar that is as real as it is intangible. The Priesthood facilitates this unity by physically administering Gospel ordinances, especially in the temple, where we are introduced to the patriarchal order of celestial marriage and are organized into eternal family units. There, we learn temporal and spiritual principles of government and make covenants to consecrate our time and talents to the Church and Kingdom, and to lend our efforts to the preparation of the Earth for the millennial reign of Jesus Christ.

The Gospel Plan is infinite and eternal in its scope, with a reach that extends across the cosmos. Without intending to do so, the theoretical physicist Steven Hawking explained both exo-biology and the existence of God, when he declared: "The quantum theory of gravity has created new possibilities, in which there is no boundary to space-time … The boundary condition of the universe is that it has no boundary." ("A Brief History of Time: From the Big Bang to Black Holes", p. 136). God is infinite "in His glory and in his power, and in his might, majesty, and dominion." (Alma 12:15).

Without the sanctifying properties of celestial light, our noblest achievements become empty shells and structures of custom and convenience only, illuminated by nothing but the flickering flames of magic and superstition. In contrast, the luminosity of the Lord liberates us from the bondage of ignorance. The warm glow that emanates from a divine fire is simply the result of a spiritual transformation in the lives of those who enjoy basking in His celestial light. It energizes our internal moral compass like the luminescent radium dial on a watch, to unerringly guide us home.

We are promised "the life and the light, the Spirit and the power, sent forth by the will of the Father through Jesus Christ, his Son." (D&C 50:27). Somewhere within the vast reaches of the galaxy, certainly there are others like us, who hope to inherit "dominion, and glory, and a kingdom", (Daniel 7:14), but they will attain the spiritual stature of their Father and our Father only if they become purified and are cleansed from all sin, by atonement, in the refining light of the Savior of worlds without end.

If we set the stylus of our moral compass squarely on this process of purification, within the circle thus scribed will be the Sacrament, the Endowment, and other ordinances that are driven by the engine of the priesthood and given vitality by the Light of the world. His power is manifest in these ordinances, and it triggers a cleansing or sanctification process that allows us to confidently ask God for our desired blessings. In such a state of humility, meekness, innocence, and holiness, we are without guile.

In the dazzling light of truth, the nature of evil that is abroad on the Earth will be plainly manifest, and we will be given "power over that spirit." (D&C 50:32). We will have the means to lay bare its true nature, "not with railing accusation" but with the measured response expected of the Lord's anointed. (D&C 50:33). If we who are privileged to be among the lucky few who make First Contact with God's extra-terrestrial children remain true to our faith, we can expect the encounter to go quite well, and possibly, something like the following scenario:

When we act in harmony with the forces that power both our physical universe and celestial glories, a divine dynamo driven by the promises of the omnipotent Lord will roar into action. The power will be given us "to overcome all things", and especially the hesitation and uncertainty that will likely accompany First Contact. (D&C 50:35-36). With the unlimited energy of the light of heaven at our disposal, however, we may very well "go forth among the" stars "and strengthen (others of God's children) by the word of exhortation." (D&C 50:37). With His power, we might very well preach the Gospel by precept and by example to those from across the galaxy, as we introduce to others of God's children the doctrines, principles, truths, and concepts pertaining to The Plan of Salvation.

Following the formalities of our interstellar introductions, we might feel privileged to be able to bring those principles into focus, illustrating in meaningful ways for our distant cousins the universal applicability of the Lord's teachings. Then, we might very well seize the opportunity to expound, or enlarge upon the principles, expanding the understanding of our extra-terrestrial congregation. Next, we might offer exhortation, instilling in them the desire to incorporate into their own lives the tenets of faith, validating their worth by encouraging ownership through personal witness and testimony.

This fanciful illustration may never be consummated in real life, but it is intriguing, nevertheless, to envision that we might one day invite others of God's children, even though they may live upon distant isles in a vast cosmic ocean, to come in out of the cold depths of space to enjoy the warm companionship of the way, the truth, the life, and the light of worlds without end. (See 2 Nephi 10:8, John 9:5 & 14:6).

Appendix One

Are We Alone in The Universe? Volume One: Chapters 1 – 24

1. Are We Alone in The Universe?..13
2. Love Letters from God..21
3. Tinker, Tailor, Soldier, Sailor..27
4. The Creation...33
5. Dancing with The Stars..39
6. The Mind of God..71
7. Is Music a Universal Language?...77
8. Habla Usted Inglés?..89
9. Heptapod Logograms and The Finger of The Lord...99
10. Synaesthesia..107
11. The Universe is a Star Nursery...115
12. If You Could Hie to Kolob..119
13. Me Transmitte Sursum, Caledoni...127
14. Does God Obey the Speed Limit?...133
15. I'm a Doctor, Not a Doormat...149
16. Travel at The Speed of Thought..155
17. The Fluidity of Time...161
18. The Q Continuum...169
19. What We Can Learn From the Q?...189
20. Is God a Carbon-based Life Form?...205
21. Is There a Prime Directive?...213
22. Our Genetic Code Has Been Scattered Across the Universe................................221
23. The Unknown Possibilities of Existence...229
24. Let There be Light..235

Are We Alone in The Universe? Volume Two: Chapters 1-27

1. Our Limiting Beliefs……………………………………………………………………13
2. Technological Traps……………………………………………………………………21
3. What is The Source of Our Power?…………………………………………………27
4. Light and Darkness……………………………………………………………………35
5. The Light of Christ in The Immensity of Space……………………………………45
6. Beings of Light…………………………………………………………………………53
7. Focus……………………………………………………………………………………59
8. Quorum Sensing: At One With The Universe………………………………………65
9. Finding Balance in a Chaotic World…………………………………………………71
10. Metaphasic Shielding…………………………………………………………………81
11. A Pale Blue Dot…………………………………………………………………………85
12. Is Anybody Out There?…………………………………………………………………91
13. Entropy in The Physical World………………………………………………………95
14. May the 4th Be With You……………………………………………………………111
15. Travel Among The Stars……………………………………………………………119
16. Tea: Earl Grey. Hot……………………………………………………………………129
17. Are We There Yet?……………………………………………………………………133
18. Pennies From Heaven………………………………………………………………141
19. The Quintessential Extra-terrestrial Being…………………………………………147
20. Wish Upon a Star……………………………………………………………………153
21. Extra-Terrestrial Voices………………………………………………………………163
22. Who is Packing Your Parachute?……………………………………………………171
23. Heaven Can Wait as We Journey to a Far Country………………………………179
24. The Ascent of Man……………………………………………………………………189
25. How Does God Get Things Done?…………………………………………………195
26. Eternal Progression in a Dynamic Universe………………………………………203
27. Higher Dimensional Realities………………………………………………………213

Appendix Two

Space: the final frontier.

These are the voyages of the starship Enterprise.

Its continuing mission: to explore strange new worlds.

To seek out new life and new civilizations.

To boldly go where no one has gone before!

Appendix Three

Are We Alone in The Universe? Scripture References

D&C 3:2	Deuteronomy 32:8	Moses 1:31
D&C 14:9	2 Kings 6:17	Moses 1:32
D&C 20:18	Ecclesiastes 3:11	Moses 1:33
D&C 29:33	Ecclesiastes 12:7	Moses 1:34
D&C 35:2	Isaiah 9:7	Moses 1:35
D&C 49:17	Isaiah 55:8-9	Moses 1:37
D&C 76:6	Jeremiah 1:5	Moses 1:38
D&C 76:7	Ezekiel 1:4	Moses 3:5
D&C 76:9	Ezekiel 1:5-6	Moses 6:36
D&C 76:10	Ezekiel 1:9	Moses 7:30
D&C 76:25	Ezekiel 1:13	Moses 7:36
D&C 76:39	Ezekiel 1:14	Abraham 2:7
D&C 76:109	Ezekiel 1:16	Abraham 3:9
D&C 78:14	Ezekiel 1:17	Abraham 3:10
D&C 88:37	Luke 1:33	Abraham 3:12
D&C 88:38	Acts 2:1-3	Abraham 3:14
D&C 88:42	Romans 8:16	Abraham 3:18
D&C 88:43	1 Corinthians 15:40	Abraham 3:22
D&C 88:44	1 Corinthians 29:10	Abraham 3:24
D&C 88:47	Ephesians 1:4	Abraham 3:26
D&C 92:23	Colossians 1:16	Abraham 4:8
D&C 92:29	2 Timothy 1:9	1 Nephi 5:11
D&C 101:32	Titus 1:2	2 Nephi 2:14
D&C 101:34	1 Peter 1:19-20	2 Nephi 11:17
D&C 121:29	Moses 1:3	Jacob 4:9
D&C 121:30	Moses 1:4	Mosiah 16:9
D&C 121:31	Moses 1:5	Alma 13:3
D&C 121:32	Moses 1:8	Alma 13:7
D&C 132:19	Moses 1:27	Alma 13:9
D&C 132:30	Moses 1:28	3 Nephi 9:15
D&C 138:56	Moses 1:29	Ether 3:16
Numbers 16:32	Moses 1:30	Ether 3:25

"God doth mot walk in crooked paths, neither doth he turn to the right hand nor to the left ... therefore, his paths are straight, and his course is one eternal round."
(D&C 3:2).

"I am Jesus Christ, the Son of the living God, who created the heavens and the earth."
(D&C 14:9).

"He created man, male and female, after his own image and in his own likeness."
(D&C 20:18).

"My works have no end, neither beginning."
(D&C 29:33).

"Listen to the voice of the Lord your God, even Alpha and Omega, the beginning and the end, whose course is one eternal round, the same today as yesterday, and forever."
(D&C 35:1-2).

"The measure of man (is) according to his creation before the world was made."
(D&C 49:17).

"Even the wonders of eternity shall they know."
(D&C 76:7).

"Their wisdom shall be great, and their understanding reach to heaven; and before them the wisdom of the wise shall perish, and the understanding of the prudent shall come to naught."
(D&C 76:9).

"For by my Spirit will I enlighten them, and by my power will I make known unto them the secrets of my will - yea, even those things which eye has not seen, nor heard, nor yet entered into the heart of man."
(D&C 76:10).

"An angel of God who was in authority in the presence of God, who rebelled against the Only Begotten Son, whom the Father loved and who was in the bosom of the Father, was thrust down from the presence of God and the Son."
(D&C 76:25).

"The Lamb ... was in the bosom of the Father before the worlds were made."
(D&C 76:39).

"We saw the glory and the inhabitants of the telestial world, that they were as innumerable as the stars in the firmament of heaven, or as the sand upon the seashore."
(D&C 76:109).

"The church ... stand(s) independent above all other creatures beneath the celestial world."
(D&C 78:14).

"There are many kingdoms; for there is no space in the which there is no kingdom; and there is no kingdom in which there is no space."
(D&C 88:37).

"Unto every kingdom is given a law; and unto every law there are certain bounds and conditions."
(D&C 88:38).

"He hath given a law unto all things, by which they move in their times and their seasons."
(D&C 88:42).

"And their courses are fixed, even the courses of the heavens and the earth, which comprehend the earth and all the planets."
(D&C 88:43).

"And they give light to each other in their times and in their seasons."
(D&C 88:44).

"All these are kingdoms, and any man who hath seen any or the least of these hath seen God moving in his majesty and power."
(D&C 88:47).

"Ye were also in the beginning with the Father."
(D&C 92:23).

"Man was also in the beginning with God. Intelligence, or the light of truth, was not created or made, neither indeed can be."
(D&C 92:29).

God "shall reveal ... things which have passed, and hidden things which no man knew, things of the earth by which it was made, and the purpose and the end thereof."
(D&C 101:32).

God shall reveal "things most precious, things that are above, and things that are beneath, things that are in the earth, and upon the earth, and in heaven."
(D&C 101:34).

"All thrones and dominions, principalities and powers, shall be revealed and set forth upon all who have endured valiantly for the gospel of Jesus Christ."
(D&C 121:29).

"If there be bounds set to the heavens or to the seas, or to the dry land, or to the sun, moon, or stars (they) shall be revealed in the days of the dispensation of the fulness of times."
(D&C 121:30).

"All the times of their revelations, all the appointed days, months, and years, and all the days of their days, months, and years, and all their glories, laws, and set times" shall be revealed.
(D&C 121:31).

"According to that which was ordained in the midst of the Council of the Eternal God of all other gods before this world was."
(D&C 121:32).

We "shall inherit thrones, kingdoms, principalities, and powers, (and) dominions, (of) all heights and depths."
(D&C 132:19).

"Abraham received promises concerning his seed," that it should "continue as innumerable as the stars; or, if ye were to count the sand upon the seashore ye could not number them."
(D&C 132:30).

"Before they were born, they, with many others, received their first lessons in the world of spirits and were prepared to come forth in the due time of the Lord."
(D&C 138:56).

"They fell upon their faces, and said, O God, the God of the spirits of all flesh" (Numbers 16:32).

"When the most High divided to the nations their inheritance, when he separated the sons of Adam, he set the bounds of the people according to the number of the children of Israel." (Deuteronomy 32:8).

"And the Lord opened the eyes of the young man; and he saw: and behold, the mountain was full of horses and chariots of fire round about Elisha." (2 Kings 6:17).

"No man can find out the work that God maketh from the beginning to the end." (Ecclesiastes 3:11).

"Then shall the dust return to the earth as it was, and the spirit shall return unto God who gave it." (Ecclesiastes 12:7).

"Of the increase of his government and peace there shall be no end." (Isaiah 9:7).

"For my thoughts are not your thoughts, neither are your ways my ways, saith the Lord. For as the heavens are higher than the earth, so are ... my thoughts than your thoughts."
(Isaiah 55:8-9).

"Before I formed thee in the belly I knew thee; and before thou camest forth out of the womb I sanctified thee, and ordained thee a prophet unto the nations."
(Jeremiah 1:5).

"I looked, and, behold, a whirlwind came out of the north, a great cloud, and a fire infolding itself."
(Ezekiel 1:4).

"Out of the midst thereof came the likeness of four living creatures ... And every one had four faces, and every one had four wings."
(Ezekiel 1:5-6).

"Their wings were joined one to another; they turned not when they went; they went every one straight forward."
(Ezekiel 1:9).

"As for the likeness of the living creatures, their appearance was like burning coals of fire, and like the appearance of lamps ... and the fire was bright, and out of the fire went forth lightning."
(Ezekiel 1:13).

"And the living creatures ran and returned as the appearance of a flash of lightning."
(Ezekiel 1:14).

"Now as I beheld the living creatures...they four had one likeness: and their appearance and their work was as it were a wheel in the middle of a wheel."
(Ezekiel 1:16).

"When they went, they went upon their four sides; and they turned not when they went."
(Ezekiel 1:17).

"He shall reign over the house of Jacob for ever; and of his kingdom there shall be no end."
(Luke 1:33).

"They were all filled ... and began to speak with other tongues as the Spirit gave them utterance."
(Acts 2:1-3).

"The Spirit itself beareth witness with our spirit, that we are the children of God."
(Romans 8:16).

"Eye hath not seen, nor ear heard, neither have entered into the heart of man, the things which God hath prepared for them that love him. But God hath revealed them unto us by his Spirit, for the Spirit searcheth all things, yea, the deep things of God."
(1 Corinthians 2:9-10).

"There are also celestial bodies, and bodies terrestrial."
(1 Corinthians 15:40).

"He hath chosen us in him before the foundation of the world."
(Ephesians 1:4).

"By him were all things created, that are in heaven, and that are in earth, visible and invisible, whether they be thrones, or dominions, or principalities, or powers: all things were created by him."
(Colossians 1:16).

God "saved us, and called us with an holy calling, not according to our works, but according to his own purpose and grace which was given us in Christ Jesus before the world began."
(2 Timothy 1:9).

"Eternal life" is that "which God, that cannot lie, promised before the world began."
(Titus 1:2).

"Christ ... was foreordained before the foundation of the world." (1 Peter 1:19-20).

"God spake unto Moses, saying: Behold, I am the Lord God Almighty, and Endless is my name; for I am without beginning of days or end of years." (Moses 1:3).

"I will show thee the workmanship of mine hands; but not all, for my works are without end." (Moses 1:4).

"No man can behold all my works." (Moses 1:5).

"Moses beheld the world and the ends thereof, and all the children of men which are, and which were created." (Moses 1:8).

"Moses cast his eyes and beheld the earth, yea, even all of it; and there was not a particle which he did not behold, discerning it by the spirit of God." (Moses 1:27).

"There was not a soul which he beheld not; and he discerned them by the Spirit of God; and their numbers were great, even numberless as the sand upon the sea shore."
(Moses 1:28).

"He beheld many lands; and each land was called earth, and there were inhabitants on the face thereof."
(Moses 1:29).

"Moses called upon God, saying: Tell me, I pray thee, why these things are so, and by what thou madest them?"
(Moses 1:30).

"And the Lord God said unto Moses: For mine own purpose have I made these things. Here is wisdom, and it remaineth in me."
(Moses 1:31).

"By the word of my power, have I created them."
(Moses 1:32).

"Worlds without number have I created."
(Moses 1:33).

"The first
man of all men
have I called Adam,
which is many."
(Moses 1:34).

"But only
an account of
this earth, and the
inhabitants thereof,
give I unto you"
(Moses 1:35).

"There are many (worlds) that
now stand and innumerable are
they unto man; but all things are
numbered unto me for they are
mine and I know them."
(Moses 1:35).

"There
are many that now
stand, and innumerable
are they unto man; but all
things are numbered unto
me, for they are mine
and I know them."
(Moses 1:35).

"The Lord God
spake unto Moses,
saying: The heavens, they
are many, and they cannot be
numbered unto man; but they
are numbered unto me,
for they are mine."
(Moses 1:37).

"As one earth shall pass away, and
the heavens thereof, even so shall
another come; and there is no
end to my works."
(Moses 1:38).

"I, the Lord God, had created all the children of men; and not yet a man to till the ground; for in heaven created I them; and there was not yet flesh upon the earth, neither in the water, neither in the air."
(Moses 3:5).

"He beheld the spirits that God had created."
(Moses 6:36).

"Were it possible that man could number the particles of the earth, yea, millions of earths like this, it would not be a beginning to the number of thy creations."
(Moses 7:30).

"I can stretch forth mine hands and hold all the creations which I have made; and mine eye can pierce them also."
(Moses 7:36).

"I dwell in heaven; the earth is my footstool; I stretch my hand over the sea, and it obeys my voice."
(Abraham 2:7).

"I cause the wind and the fire to be my chariot; I say to the mountains - Depart hence, and behold, they are taken away by a whirlwind, in an instant, suddenly."
(Abraham 2:7).

"There shall be the reckoning of the time of one planet above another, until thou come nigh unto Kolob."
(Abraham 3:9).

"It is given unto thee to know the set time of all the stars that are set to give light, until thou come near unto the throne of God."
(Abraham 3:10).

"And he put his hand upon mine eyes, and I saw those things which his hands had made, which were many; and they multiplied before mine eyes, and I could not see the end thereof."
(Abraham 3:12).

"I will multiply thee, and thy seed after thee, like unto these; and if thou canst count the number of sands so shall be the number of thy seeds."
(Abraham 3:14).

"They existed before, they shall have no end, they shall exist after, for they are ... eternal."
(Abraham 3:18).

"The Lord has shown unto me, Abraham, the intelligences that were organized before the world was."
(Abraham 3:22).

"There stood one among them that was like unto God, and he said unto those who were with him: we will go down, for there is space there, and we will take of these materials, and we will make an earth whereon these may dwell." (Abraham 3:24).

"And they who keep their first estate shall be added upon ... and they who keep their second estate shall have glory added upon their heads for ever and ever." (Abraham 3:26).

"And the Gods called the expanse, Heaven." (Abraham 4:8).

"The five books of Moses ... gave an account of the creation of the world, and also of Adam and Eve, who were our first parents." (1 Nephi 5:11).

"There is a God, and he hath created all things, both the heavens and the earth, and all things that in them are." (2 Nephi 2:14).

"If there be no Christ there be no God; and if there be no God we are not, for there could have been no creation." (2 Nephi 11:7).

"By the power of his word, man came upon the face of the earth, which earth was created by the power of His word."
(Jacob 4:9).

"He is the light and the life of the world; yea, a light that is endless, that can never be darkened."
(Mosiah 16:9).

"This is the manner after which they were ordained – being called and prepared from the foundation of the world according to the foreknowledge of God."
(Alma 13:3).

"This high priesthood ... being without beginning of days or end of years, being prepared from eternity to all eternity according to his foreknowledge of all things."
(Alma 13:7).

"High priests (are) after the order of the Son, the Only Begotten of the Father, who is without beginning of days or end of years."
(Alma 13:9).

"I am Jesus Christ the Son of God. I created the heavens and the earth, and all things that in them are."
(3 Nephi 9:15).

"Man have I created after the body of my spirit."
(Ether 3:16).

"He showed unto the brother of Jared all the inhabitants of the earth which had been, and also all that would be; and he withheld them not from his sight, even unto the ends of the earth."
(Ether 3:25).

hotter than hell

Appendix Four

hotter than hell

Is Heaven Hotter Than Hell?

Today, interstellar space has an average temperature of 2.7° Kelvin, which is minus 455° Fahrenheit, or minus 270.45° Celsius. This is the temperature of the cosmic microwave background radiation, that is the measurable residual heat of the universe. It is left over from the Big Bang. During creation, at the moment of singularity, the volume of space was zero and its density was infinite. One could say that the curvature of spacetime was also infinite.

No-one really knows for sure what happened during the first second after Creation, but within a trillionth of a second, scientists postulate that the universe grew to nearly one octillion times its original measurable size. This "inflation" was followed by more gradual expansion, during which the universe cooled enough for protons and neutron to stick together as they emerged from the primordial quark soup. Also known as quark-gluon plasma, the soup was a liquid-like material formed out of the most basic building blocks of matter, quarks and gluons.

Initially, the temperature of the proto universe was 100 trillion degrees Celsius. After just 100 seconds, the temperature had cooled to a balmy one billion degrees Celsius. Over 13.7 billion years, things have cooled down to the relatively stable temperature of 2.7° Kelvin.

In our day, with the fireworks of the Big Bang a thing of the past, interstellar explorers from plant Earth will have to deal with a number of challenging issues relating to human physiology. Our dread of the cold vacuum of space, however, might not be one of them. It is a misconception that space is cold; in fact, it has no temperature at all, because in thermodynamic terms, temperature is a function of heat energy as it relates to mass, and space has none. Heat transfer in its vacuum cannot take place by either conduction or convection. In space, it can only occur by the process of thermal radiation, which is not very efficient. A space traveler might feel the warmth of the Sun's (or another star's) rays, when directly exposed. Or, they might feel cool, when shielded from those rays. But, with proper atmospheric protection, they are not going to freeze.

Even if you were "spaced", or unceremoniously ejected from the cargo bay of a starship by a malevolent entity, you would not freeze instantly, because heat transfer cannot rapidly occur by radiation alone. (See: "The Expanse" television series).

Assuming space travelers can maintain body temperature and are able to avoid being spaced by their traveling companions, there are other things to worry about. The effects of microgravity will affect human organ systems and compromise physical fitness. Because our cardiovascular systems are designed to work against gravity, the redistribution of body fluids due to the weightlessness of space will occur within a few moments, Other cardiovascular phenomena resulting from travel in the weightlessness of space include a reduction in blood volume, red blood cell quantity, and cardiac output. These phenomena might pose little problem while in space but will be problematic following a return to the terrestrial force of 1 G. gravity.

The lack of gravity can cause other more daunting problems, such as muscle atrophy (up to 50% loss of muscle mass), demineralization of bones, and decreased bone density. Calcium that is excreted in urine at elevated levels will increase the risk of kidney stones. The vestibular system might fail, leading to imbalance, nausea, disorientation, and decreased motor coordination.

Without the Earth's atmosphere to act as a shield from solar radiation, space travelers will be exposed to dangerous ultraviolet rays. An unprotected human in space will suffer sunburn within seconds of exposure. With current technology, even a two-year mission to our nearest planetary neighbor (Mars) would expose human astronauts to dangerous levels of radiation. Protective shielding and other technologies that address the physiological limitations of the human body have been designed by NASA, but high energy ionizing radiation continues to pose challenges for engineers designing suitable habitat and clothing options for astronauts who may be asked to dance with the stars for extended periods of time.

One of the things that we do not need to grapple with, at least in the immediate future assuming we stay out of harm's way, is the temperature of heaven and hell. Hopefully, we will one day carry out manned missions to our planetary neighbors while navigating well between those two extremes. Nevertheless, because each of us will eventually make the journey to that undiscovered country from whose bourne no traveler has ever returned, it might be worthwhile to know if we can anticipate the atmosphere of heaven, or hell for that matter, to be pleasant or not. In 1972, scientists attempted to answer that question, once and for all. In an article in "Applied Optics" (11:14) they stated that heaven must be hotter than hell. The paper noted that Revelation 21:8 describes a lake in hell "which burneth with fire and brimstone". For there to be such a lake, the authors reasoned, Hell's temperature must be below the boiling point of sulphur, which is $444.6\frac{1}{4}°$ C. At a higher temperature, the lake's contents would simply boil away. Meanwhile, they noted that the Old Testament describes the conditions in heaven, where "the light of the Moon shall be as the light of the Sun, and the light of the Sun shall be sevenfold, as the light of seven days." (Isaiah 30:26).

The authors of the article interpreted this to mean that if "the light of seven days" is a metaphorical way of describing heaven, then we can determine that heaven receives from the moon as much radiation as we do from the sun, and in addition, seven times seven as much as the earth does from the sun, or fifty times as much in all.

Since the light we receive from the moon is just a ten-thousandth of the light we receive from the sun, we can ignore it as insignificant. The radiation falling on heaven will heat it to the point where the heat lost is just equal to the heat received, maintaining a state of equilibrium. Therefore, heaven receives and loses 50 times as much heat as does the earth, by radiation. Using the Stefan-Boltzmann Fourth Power Law for Radiation, they calculated the temperature of heaven to be a sweltering $525\frac{1}{4}°$ C. Heaven, they concluded, really is hotter than hell!

But wait! A rebuttal letter that was published in the magazine "Physics Today" argued that the "Applied Optics" authors significantly misinterpreted the data in the Isaiah passage, wrongly multiplying seven by seven to make the illumination in heaven forty-nine times as bright as that which we experience on earth. They pointed out that theologians have long argued that only a single factor of seven was intended by Isaiah. Consequently, they recalculated heaven's temperature to be only 231° C., still blisteringly hot, but probably cooler than hell.

The question: "Is heaven hotter than hell?" was thus left open to debate. The definitive answer may elude us, until we make the journey to one or the other, but for now we can draw upon the experience of a thermodynamics professor who gave his graduate students a take home exam. It had only one question: "Is Hell exothermic (does it give off heat) or is it endothermic (does it absorb heat?)" The answer was to be supported with a proof. Most of the students based their proofs on Boyle's Law: Gas cools off when it expands, and it heats up when it is compressed.

But one visionary student wrote the following: "First, we need to understand how the mass of hell is changing over time. So, we need to know the rate at which souls are arriving in hell, as well as the rate at which they are leaving. I think we can safely assume that once a soul gets to hell, it stays there. As for how many souls are entering hell, consider that most religions state that if you do not believe as they do, you will go to hell. So, we can assume that nearly everyone is going to hell. As a matter of fact, we can expect the number of souls in hell to be increase exponentially. This leaves two possibilities, according to Boyle's Law.

#1. If hell is expanding slower than the rate at which souls enter hell, then the temperature and pressure in hell will increase, until all hell breaks loose.

#2. If hell is expanding at a rate that is faster than the increase in the number of souls in hell, then the temperature and pressure will drop, until hell freezes over.

"So, which is it?" he asked. "If we accept the postulate given to me by Jackie Larson during my freshman year that "it will be a cold night in hell before I go out with you," and if we consider the fact that we never dated, then #2 cannot be true, and so hell must be exothermic." (He got the only A.).

About The Author

Phil Hudson and his wife Jan have 7 children and 25 grandchildren. They enjoy spending time with their family at their cabin nestled in the Selkirk Mountains, on the shore of Priest Lake, the crown jewel of North Idaho. Phil had a successful dental practice in Spokane, Washington for 43 years, before retiring in 2015. He has an eclectic mix of hobbies and enjoys the out of doors. He always finds time, however, to record his thoughts on his laptop, and understands Isaac Asimov's response when he was asked: "If you knew that you had only 10 minutes left to live, what would you do?" He answered: "I'd type faster."

Phil received the inspiration to write this book while he and Jan were serving as missionaries for The Church of Jesus Christ of Latter-day Saints, in the Kingdom of Tonga. While there, they celebrated their 50th wedding anniversary.

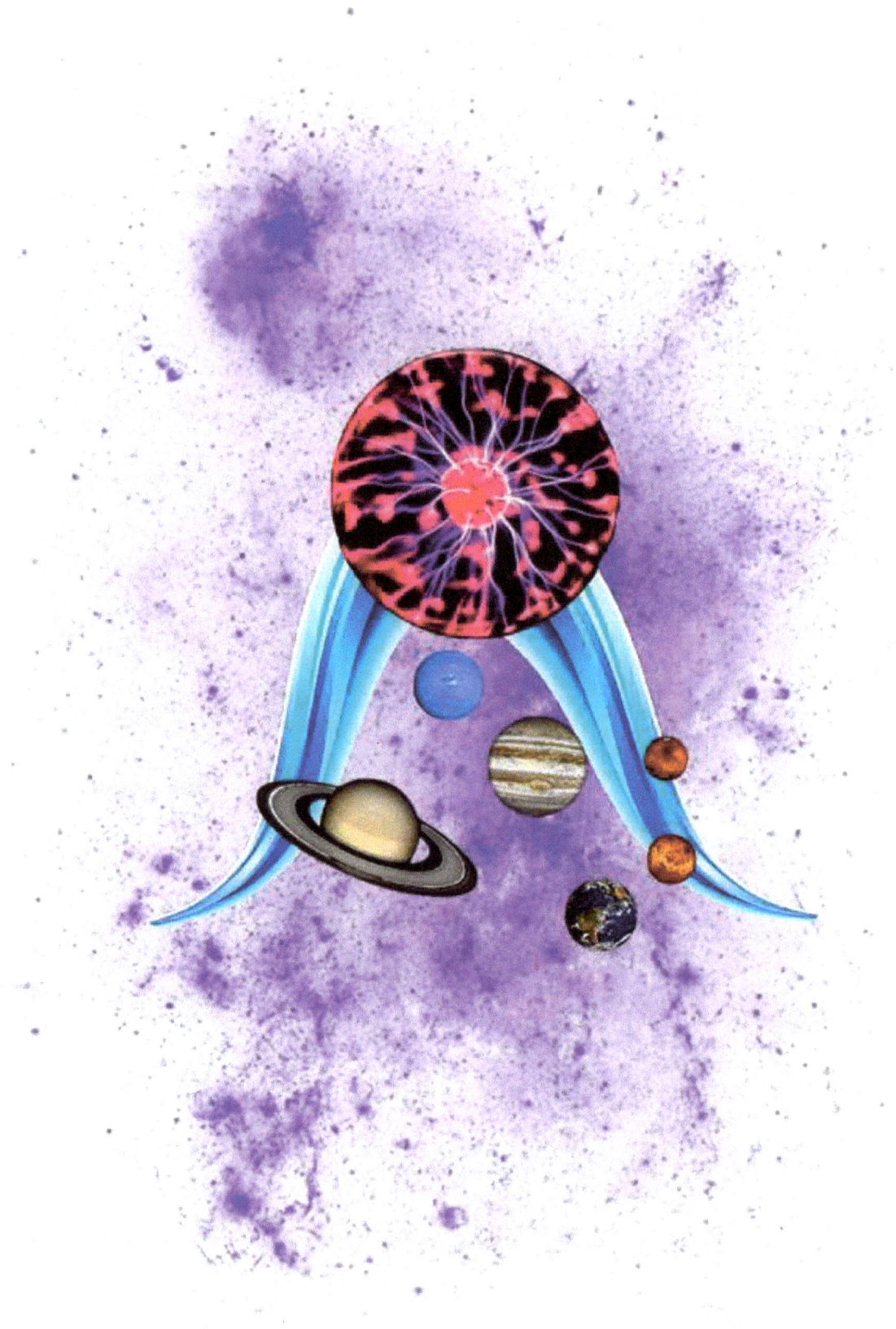

By The Author

Essays

- Volume 1 - Spray from The Ocean of Thought
- Volume 2 - Ripples on a Pond
- Volume 3 - Serendipitous Meanderings
- Volume 4 - Presents of Mind
- Volume 5 - Mental Floss
- Volume 6 - Fitness Training for the Mind and Spirit

First Principles and Ordinances Series

- Faith - Our Hearts are Changed
- Repentance - A Broken Heart and a Contrite Spirit
- Baptism - One Hundred and One Reasons Why We Are Baptized
- Holy Ghost - That We Might Have His Spirit to Be With Us
- Sacrament - This Do in Remembrance of Me

Minute Musings - Spontaneous Combustions of Thought

- Volume One
- Volume Two
- Volume Three

Book of Mormon Commentary

- Volume One - Born in The Wilderness
- Volume Two - Voices from The Dust
- Volume Three - Journey to Cumorah

Calendars

- In His Own Words - Discovering William Tyndale
- As I Think About the Savior
- Daily Inspiration from Scriptural Symbols

A Thought for Each Day of the Year

- Faith
- Repentance
- Baptism
- The Holy Ghost
- The Sacrament

    Life's Greatest Questions
    Revelation
    The Atonement
    The House of the Lord
    The Plan of Salvation
    The Sabbath

Doctrine & Covenants Commentary

    Volume One - Sections 1 - 34
    Volume Two - Sections 35 - 57

Doctrinal Themes

    Are Christians Mormon? - Volume One
    Are Christians Mormon? - Volume Two
    Are We Alone in The Universe? – Volume One
    Are We Alone in The Universe? – Volume Two
    Christmas is The Season When …
    Dancing With the Stars – Volume One
    Dancing With the Stars – Volume Two
    Dancing With the Stars – Volume Three
    Dancing With the Stars – Volume Four
    Dentistry in The Scriptures
    Gratitude
    Hebrew Poetry
    Hiding in Plain Sight
    One Hundred Questions Answered by The Book of Mormon
    The Highways and Byways of Life – Volume One
    The Highways and Byways of Life – Volume Two
    The Highways and Byways of Life – Volume Three
    The House of The Lord
    Without the Book of Mormon
    Writing on Metal Plates

Children's Books

    Book of Mormon Hiking Song – Volume One
    Book of Mormon Hiking Song – Volume Two
    Book of Mormon Hiking Song – Volume Three
    Happy Birthday
    Muddy, Muddy
    The Hiawatha Trail - An Allegory
    The Little Princess

The Parable of The Pencil
The Strange Tale of Huckleberry Henry
The Thirteen Articles of Faith

Professional Publications

Diode Laser Soft Tissue Surgery – Volume One
Diode Laser Soft Tissue Surgery – Volume Two
Diode Laser Soft Tissue Surgery – Volume Three

These, and other titles, are available from online retailers.

Quid Magis Possum Dicire?

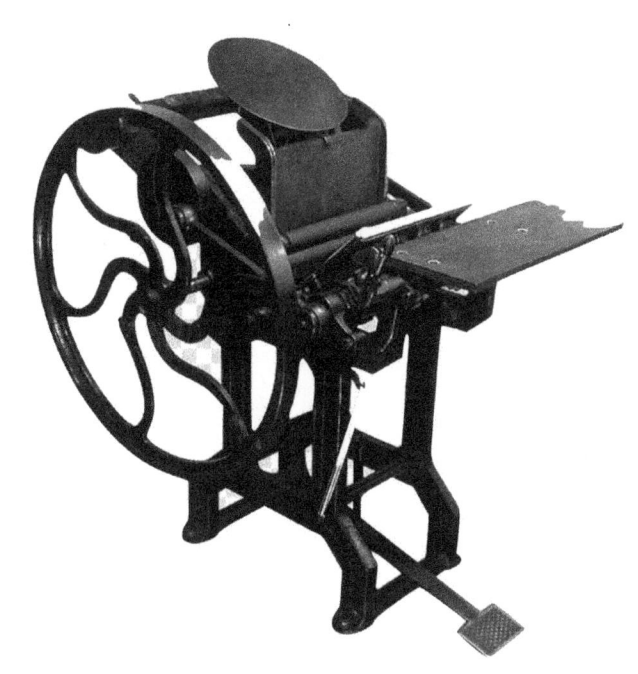

www.ingramcontent.com/pod-product-compliance
Lightning Source LLC
Chambersburg PA
CBHW041236240426
43673CB00011B/355